CREATIVE CHEMISTRY

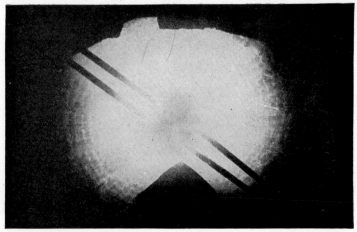

Courtesy of E. I. du Pont de Nemours Co.

BURNING AIR IN A BIRKELAND-EYDE FURNACE AT THE DU PONT PLANT

An electric arc consuming about 4000 horse-power of energy is passing between the U-shaped electrodes which are made of copper tube cooled by an internal current of water. On the sides of the chamber are seen the openings through which the air passes impinging directly on both sides of the surface of the disk of flame. This flame is approximately seven feet in diameter and appears to be continuous although an alternating current of fifty cycles a second is used. The electric arc is spread into this disk flame by the repellent power of an electro-magnet the pointed pole of which is seen at the bottom of the picture. Under this intense heat a part of the nitrogen and oxygen of the air combine to form oxides of nitrogen which when dissolved in water form the nitric acid used in explosives.

Courtesy of E. I. du Pont de Nemours Co.

A BATTERY OF BIRKELAND-EYDE FURNACES FOR THE FIXATION OF NITROGEN AT THE DU PONT PLANT

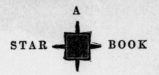

A STAR BOOK

CREATIVE CHEMISTRY

DESCRIPTIVE OF RECENT ACHIEVEMENTS
IN THE CHEMICAL INDUSTRIES

BY
EDWIN E. SLOSSON
M.S., Ph.D.

ILLUSTRATED

GARDEN CITY PUBLISHING CO., INC.
GARDEN CITY, NEW YORK

Copyright, 1919, by
THE CENTURY CO.

Copyright, 1917, 1918, 1919, by
THE INDEPENDENT CORPORATION

TO MY FIRST TEACHER

PROFESSOR E. H. S. BAILEY
OF THE UNIVERSITY OF KANSAS

AND MY LAST TEACHER

PROFESSOR JULIUS STIEGLITZ
OF THE UNIVERSITY OF CHICAGO

THIS VOLUME IS GRATEFULLY
DEDICATED

CONTENTS

		PAGE
I	Three Periods of Progress	3
II	Nitrogen	14
III	Feeding the Soil	37
IV	Coal-Tar Colors	60
V	Synthetic Perfumes and Flavors	93
VI	Cellulose	110
VII	Synthetic Plastics	128
VIII	The Race for Rubber	145
IX	The Rival Sugars	164
X	What Comes from Corn	181
XI	Solidified Sunshine	196
XII	Fighting with Fumes	218
XIII	Products of the Electric Furnace	236
XIV	Metals, Old and New	263
	Reading References	297
	Index	309

A CARD OF THANKS

This book originated in a series of articles prepared for *The Independent* in 1917-18 for the purpose of interesting the general reader in the recent achievements of industrial chemistry and providing supplementary reading for students of chemistry in colleges and high schools. I am indebted to Hamilton Holt, editor of *The Independent*, and to Karl V. S. Howland, its publisher, for stimulus and opportunity to undertake the writing of these pages and for the privilege of reprinting them in this form.

In gathering the material for this volume I have received the kindly aid of so many companies and individuals that it is impossible to thank them all but I must at least mention as those to whom I am especially grateful for information, advice and criticism: Thomas H. Norton of the Department of Commerce; Dr. Bernhard C. Hesse; H. S. Bailey of the Department of Agriculture; Professor Julius Stieglitz of the University of Chicago; L. E. Edgar of the Du Pont de Nemours Company; Milton Whitney of the U. S. Bureau of Soils; Dr. H. N. McCoy; K. F. Kellerman of the Bureau of Plant Industry.

E. E. S.

INTRODUCTION

By JULIUS STIEGLITZ

Formerly President of the American Chemical Society, Professor of Chemistry in The University of Chicago

The recent war as never before in the history of the world brought to the nations of the earth a realization of the vital place which the science of chemistry holds in the development of the resources of a nation. Some of the most picturesque features of this awakening reached the great public through the press. Thus, the adventurous trips of the *Deutschland* with its cargoes of concentrated aniline dyes, valued at millions of dollars, emphasized as no other incident our former dependence upon Germany for these products of her chemical industries.

The public read, too, that her chemists saved Germany from an early disastrous defeat, both in the field of military operations and in the matter of economic supplies: unquestionably, without the tremendous expansion of her plants for the production of nitrates and ammonia from the air by the processes of Haber, Ostwald and others of her great chemists, the war would have ended in 1915, or early in 1916, from exhaustion of Germany's supplies of nitrate explosives, if not indeed from exhaustion of her food supplies as a consequence of the lack of nitrate and ammonia fertilizer for her fields. Inventions of substitutes for cotton, copper, rubber, wool and many other basic needs have been reported.

INTRODUCTION

These feats of chemistry, performed under the stress of dire necessity, have, no doubt, excited the wonder and interest of our public. It is far more important at this time, however, when both for war and for peace needs, the resources of our country are strained to the utmost, that the public should awaken to a clear realization of what this science of chemistry really means for mankind, to the realization that its wizardry permeates the whole life of the nation as a vitalizing, protective and constructive agent very much in the same way as our blood, coursing through our veins and arteries, carries the constructive, defensive and life-bringing materials to every organ in the body.

If the layman will but understand that chemistry is the fundamental *science of the transformation of matter,* he will readily accept the validity of this sweeping assertion: he will realize, for instance, why exactly the same fundamental laws of the science apply to, and make possible scientific control of, such widely divergent national industries as agriculture and steel manufacturing. It governs the transformation of the salts, minerals and humus of our fields and the components of the air into corn, wheat, cotton and the innumerable other products of the soil; it governs no less the transformation of crude ores into steel and alloys, which, with the cunning born of chemical knowledge, may be given practically any conceivable quality of hardness, elasticity, toughness or strength. And exactly the same thing may be said of the hundreds of national activities that lie between the two extremes of agriculture and steel manufacture!

Moreover, the domain of the science of the transfor-

INTRODUCTION

mation of matter includes even life itself as its loftiest phase: from our birth to our return to dust the laws of chemistry are the controlling laws of life, health, disease and death, and the ever clearer recognition of this relation is the strongest force that is raising medicine from the uncertain realm of an art to the safer sphere of an exact science. To many scientific minds it has even become evident that those most wonderful facts of life, heredity and character, must find their final explanation in the chemical composition of the components of life producing, germinal protoplasm: mere form and shape are no longer supreme but are relegated to their proper place as the housing only of the living matter which functions chemically.

It must be quite obvious now why thoughtful men are insisting that the public should be awakened to a broad realization of the significance of the science of chemistry for its national life.

It is a difficult science in its details, because it has found that it can best interpret the visible phenomena of the material world on the basis of the conception of invisible minute material atoms and molecules, each a world in itself, whose properties may be nevertheless accurately deduced by a rigorous logic controlling the highest type of scientific imagination. But a layman is interested in the wonders of great bridges and of monumental buildings without feeling the need of inquiring into the painfully minute and extended calculations of the engineer and architect of the strains and stresses to which every pin and every bar of the great bridge and every bit of stone, every foot of arch in a monumental edifice, will be exposed. So the public may

INTRODUCTION

understand and appreciate with the keenest interest the results of chemical effort without the need of instruction in the intricacies of our logic, of our dealings with our minute, invisible particles.

The whole nation's welfare demands, indeed, that our public be enlightened in the matter of the relation of chemistry to our national life. Thus, if our commerce and our industries are to survive the terrific competition that must follow the reëstablishment of peace, our public must insist that its representatives in Congress preserve that independence in chemical manufacturing which the war has forced upon us in the matter of dyes, of numberless invaluable remedies to cure and relieve suffering; in the matter, too, of hundreds of chemicals, which our industries need for their successful existence.

Unless we are independent in these fields, how easily might an unscrupulous competing nation do us untold harm by the mere device, for instance, of delaying supplies, or by sending inferior materials to this country or by underselling our chemical manufacturers and, after the destruction of our chemical independence, handicapping our industries as they were in the first year or two of the great war! This is not a mere possibility created by the imagination, for our economic history contains instance after instance of the purposeful undermining and destruction of our industries in finer chemicals, dyes and drugs by foreign interests bent on preserving their monopoly. If one recalls that through control, for instance, of dyes by a competing nation, control is in fact also established over products, valued in the hundreds of millions of dollars, in

INTRODUCTION

which dyes enter as an essential factor, one may realize indeed the tremendous industrial and commercial power which is controlled by the single lever—chemical dyes. Of even more vital moment is chemistry in the domain of health: the pitiful calls of our hospitals for local anesthetics to alleviate suffering on the operating table, the frantic appeals for the hypnotic that soothes the epileptic and staves off his seizure, the almost furious demands for remedy after remedy, that came in the early years of the war, are still ringing in the hearts of many of us. No wonder that our small army of chemists is grimly determined not to give up the independence in chemistry which war has achieved for us! Only a widely enlightened public, however, can insure the permanence of what farseeing men have started to accomplish in developing the power of chemistry through research in every domain which chemistry touches.

The general public should realize that in the support of great chemical research laboratories of universities and technical schools it will be sustaining important centers from which the science which improves products, abolishes waste, establishes new industries and preserves life, may reach out helpfully into all the activities of our great nation, that are dependent on the transformation of matter.

The public is to be congratulated upon the fact that the writer of the present volume is better qualified than any other man in the country to bring home to his readers some of the great results of modern chemical activity as well as some of the big problems which must continue to engage the attention of our chemists. Dr.

INTRODUCTION

Slosson has indeed the unique quality of combining an exact and intimate knowledge of chemistry with the exquisite clarity and pointedness of expression of a born writer.

We have here an exposition by a master mind, an exposition shorn of the terrifying and obscuring technicalities of the lecture room, that will be as absorbing reading as any thrilling romance. For the story of scientific achievement is the greatest epic the world has ever known, and like the great national epics of bygone ages, should quicken the life of the nation by a realization of its powers and a picture of its possibilities.

CREATIVE CHEMISTRY

La Chimie posséde cette faculté créatrice à un degré plus éminent que les autres sciences, parce qu'elle pénètre plus profondément et atteint jusqu'aux éléments naturels des êtres.
—*Berthelot.*

CREATIVE CHEMISTRY

1

THREE PERIODS OF PROGRESS

The story of Robinson Crusoe is an allegory of human history. Man is a castaway upon a desert planet, isolated from other inhabited worlds—if there be any such—by millions of miles of untraversable space. He is absolutely dependent upon his own exertions, for this world of his, as Wells says, has no imports except meteorites and no exports of any kind. Man has no wrecked ship from a former civilization to draw upon for tools and weapons, but must utilize as best he may such raw materials as he can find. In this conquest of nature by man there are three stages distinguishable:

1. The Appropriative Period
2. The Adaptive Period
3. The Creative Period

These eras overlap, and the human race, or rather its vanguard, civilized man, may be passing into the third stage in one field of human endeavor while still lingering in the second or first in some other respect. But in any particular line this sequence is followed. The primitive man picks up whatever he can find available for his use. His successor in the next stage of culture shapes and develops this crude instrument until it becomes more suitable for his purpose. But in the course of time man often finds that he can make something new which is better than anything in nature or naturally produced. The savage discovers. The barbarian improves. The civilized man invents. The first finds. The second fashions. The third fabricates.

The primitive man was a troglodyte. He sought shelter in any cave or crevice that he could find. Later he dug it out to make it more roomy and piled up stones at the entrance to keep out the wild beasts. This artificial barricade, this false façade, was gradually extended and solidified until finally man could build a cave for himself anywhere in the open field from stones he quarried out of the hill. But man was not content with such materials and now puts up a building which may be composed of steel, brick, terra cotta, glass, concrete and plaster, none of which materials are to be found in nature.

The untutored savage might cross a stream astride a floating tree trunk. By and by it occurred to him to sit inside the log instead of on it, so he hollowed it out with fire or flint. Later, much later, he constructed an ocean liner.

Cain, or whoever it was first slew his brother man, made use of a stone or stick. Afterward it was found a better weapon could be made by tying the stone to the end of the stick, and as murder developed into a fine art the stick was converted into the bow and this into the catapult and finally into the cannon, while the stone was developed into the high explosive projectile.

The first music to soothe the savage breast was the

soughing of the wind through the trees. Then strings were stretched across a crevice for the wind to play upon and there was the Æolian harp. The second stage was entered when Hermes strung the tortoise shell and plucked it with his fingers and when Athena, raising the wind from her own lungs, forced it through a hollow reed. From these beginnings we have the organ and the orchestra, producing such sounds as nothing in nature can equal.

The first idol was doubtless a meteorite fallen from heaven or a fulgurite or concretion picked up from the sand, bearing some slight resemblance to a human being. Later man made gods in his own image, and so sculpture and painting grew until now the creations of futuristic art could be worshiped—if one wanted to—without violation of the second commandment, for they are not the likeness of anything that is in heaven above or that is in the earth beneath or that is in the water under the earth.

In the textile industry the same development is observable. The primitive man used the skins of animals he had slain to protect his own skin. In the course of time he—or more probably his wife, for it is to the women rather than to the men that we owe the early steps in the arts and sciences—fastened leaves together or pounded out bark to make garments. Later fibers were plucked from the sheepskin, the cocoon and the cotton-ball, twisted together and woven into cloth. Nowadays it is possible to make a complete suit of clothes, from hat to shoes, of any desirable texture, form and color, and not include any substance to be found in nature. The first metals available were those found free in nature such as gold and copper. In a later age it was found possible to extract iron from its ores and today we have artificial alloys made of multifarious combinations of rare metals. The medicine man dosed his patients with decoctions of such roots and herbs as had a bad taste or queer look. The pharmacist discovered how to extract from these their medicinal principle such as morphine, quinine and cocaine, and the creative chemist has discovered how to make innumerable drugs adapted to specific diseases and individual idiosyncrasies.

In the later or creative stages we enter the domain of chemistry, for it is the chemist alone who possesses the power of reducing a substance to its constituent atoms and from them producing substances entirely new. But the chemist has been slow to realize his unique power and the world has been still slower to utilize his invaluable services. Until recently indeed the leaders of chemical science expressly disclaimed what should have been their proudest boast. The French chemist Lavoisier in 1793 defined chemistry as "the science of analysis." The German chemist Gerhardt in 1844 said: "I have demonstrated that the chemist works in opposition to living nature, that he burns, destroys, analyzes, that the vital force alone operates by synthesis, that it reconstructs the edifice torn down by the chemical forces."

It is quite true that chemists up to the middle of the last century were so absorbed in the destructive side of their science that they were blind to the constructive side of it. In this respect they were less prescient than their contemned predecessors, the alchemists, who, fool-

THREE PERIODS OF PROGRESS

ish and pretentious as they were, aspired at least to the formation of something new.

It was, I think, the French chemist Berthelot who first clearly perceived the double aspect of chemistry, for he defined it as "the science of analysis *and synthesis*," of taking apart and of putting together. The motto of chemistry, as of all the empirical sciences, is *savoir c'est pouvoir*, to know in order to do. This is the pragmatic test of all useful knowledge. Berthelot goes on to say:

> Chemistry creates its object. This creative faculty, comparable to that of art itself, distinguishes it essentially from the natural and historical sciences. . . . These sciences do not control their object. Thus they are too often condemned to an eternal impotence in the search for truth of which they must content themselves with possessing some few and often uncertain fragments. On the contrary, the experimental sciences have the power to realize their conjectures. . . . What they dream of that they can manifest in actuality. . . .
>
> Chemistry possesses this creative faculty to a more eminent degree than the other sciences because it penetrates more profoundly and attains even to the natural elements of existences.

Since Berthelot's time, that is, within the last fifty years, chemistry has won its chief triumphs in the field of synthesis. Organic chemistry, that is, the chemistry of the carbon compounds, so called because it was formerly assumed, as Gerhardt says, that they could only be formed by "vital force" of organized plants and animals, has taken a development far overshadowing inorganic chemistry, or the chemistry of mineral substances. Chemists have prepared or know how to prepare hundreds of thousands of such "organic compounds," few of which occur in the natural world.

But this conception of chemistry is yet far from having been accepted by the world at large. This was brought forcibly to my attention during the publication of these chapters in "The Independent" by various letters, raising such objections as the following:

> When you say in your article on "What Comes from Coal Tar" that "Art can go ahead of nature in the dyestuff business" you have doubtless for the moment allowed your enthusiasm to sweep you away from the moorings of reason. Shakespeare, anticipating you and your "Creative Chemistry," has shown the utter untenableness of your position:
>> Nature is made better by no mean,
>> But nature makes that mean: so o'er that art,
>> Which, you say, adds to nature, is an art
>> That nature makes.
>
> How can you say that art surpasses nature when you know very well that nothing man is able to make can in any way equal the perfection of all nature's products?
>
> It is blasphemous of you to claim that man can improve the works of God as they appear in nature. Only the Creator can create. Man only imitates, destroys or defiles God's handiwork.

No, it was not in momentary absence of mind that I claimed that man could improve upon nature in the making of dyes. I not only said it, but I proved it. I not only proved it, but I can back it up. I will give a million dollars to anybody finding in nature dyestuffs as numerous, varied, brilliant, pure and cheap as those that are manufactured in the laboratory. I haven't that amount of money with me at the moment, but the dyers would be glad to put it up for the discovery of a satisfactory natural source for their tinctorial materials. This is not an opinion of mine but a matter of fact, not to be decided by Shakespeare, who was not acquainted with the aniline products.

Shakespeare in the passage quoted is indulging in his favorite amusement of a play upon words. There is a possible and a proper sense of the word "nature" that makes it include everything except the supernatural. Therefore man and all his works belong to the realm of nature. A tenement house in this sense is as "natural" as a bird's nest, a peapod or a crystal.

But such a wide extension of the term destroys its distinctive value. It is more convenient and quite as correct to use "nature" as I have used it, in contradistinction to "art," meaning by the former the products of the mineral, vegetable and animal kingdoms, excluding the designs, inventions and constructions of man which we call "art."

We cannot, in a general and abstract fashion, say which is superior, art or nature, because it all depends on the point of view. The worm loves a rotten log into which he can bore. Man prefers a steel cabinet into which the worm cannot bore. If man cannot improve upon nature he has no motive for making anything. Artificial products are therefore superior to natural products as measured by man's convenience, otherwise they would have no reason for existence.

Science and Christianity are at one in abhorring the natural man and calling upon the civilized man to fight and subdue him. The conquest of nature, not the imitation of nature, is the whole duty of man. Metchnikoff and St. Paul unite in criticizing the body we were born with. St. Augustine and Huxley are in agreement as to the eternal conflict between man and nature. In his Romanes lecture on "Evolution and Ethics" Huxley said: "The ethical progress of society depends, not on imitating the cosmic process, still less on running away from it, but on combating it," and again: "The history of civilization details the steps by which man has succeeded in building up an artificial world within the cosmos."

There speaks the true evolutionist, whose one desire is to get away from nature as fast and far as possible. Imitate Nature? Yes, when we cannot improve upon her. Admire Nature? Possibly, but be not blinded to her defects. Learn from Nature? We should sit humbly at her feet until we can stand erect and go our own way. Love Nature? Never! She is our treacherous and unsleeping foe, ever to be feared and watched and circumvented, for at any moment and in spite of all our vigilance she may wipe out the human race by famine, pestilence or earthquake and within a few centuries obliterate every trace of its achievement. The wild beasts that man has kept at bay for a few centuries will in the end invade his palaces: the moss will envelop his walls and the lichen disrupt them. The clam may survive man by as many millennia as it preceded him. In the ultimate devolution of the world animal life will disappear before vegetable, the higher plants will be killed off before the lower, and finally the three kingdoms of nature will be reduced to one, the mineral. Civilized man, enthroned in his citadel and defended by all the forces of nature that he has brought under his control, is after all in the same situation as a savage, shivering in the darkness beside his fire, listening to the pad of predatory feet, the rustle of serpents and the cry of birds of prey, knowing that only the fire keeps his enemies off, but knowing too that every stick he lays on the fire lessens his fuel supply and hastens the inevitable time when the beasts of the jungle will make their fatal rush.

Chaos is the "natural" state of the universe. Cosmos is the rare and temporary exception. Of all the million spheres this is apparently the only one habitable and of this only a small part—the reader may draw the boundaries to suit himself—can be called civilized. Anarchy is the natural state of the human race. It prevailed exclusively all over the world up to some five thousand years ago, since which a few peoples have for a time succeeded in establishing a certain degree of peace and order. This, however, can be maintained only by strenuous and persistent efforts, for society tends naturally to sink into the chaos out of which it has arisen.

It is only by overcoming nature that man can rise. The sole salvation for the human race lies in the removal of the primal curse, the sentence of hard labor for life that was imposed on man as he left Paradise. Some folks are trying to elevate the laboring classes; some are trying to keep them down. The scientist has a more radical remedy; he wants to annihilate the laboring classes by abolishing labor. There is no longer any need for human labor in the sense of personal toil, for the physical energy necessary to accomplish all kinds of work may be obtained from external sources and it can be directed and controlled without extreme exertion. Man's first effort in this direction was to throw part of his burden upon the horse and ox or upon other men. But within the last century it has been discovered that neither human nor animal servitude is necessary to give man leisure for the higher life, for by means of the machine he can do the work of giants without exhaustion. But the introduction of machines, like every other step of human progress, met with the most violent opposition from those it was to benefit. "Smash 'em!" cried the workingman. "Smash 'em!" cried the poet. "Smash 'em!" cried the artist. "Smash 'em!" cried the theologian. "Smash 'em!" cried the magistrate. This opposition yet lingers and every new invention, especially in chemistry, is greeted with general distrust and often with legislative prohibition.

Man is the tool-using animal, and the machine, that is, the power-driven tool, is his peculiar achievement. It is purely a creation of the human mind. The wheel, its essential feature, does not exist in nature. The lever, with its to-and-fro motion, we find in the limbs of all animals, but the continuous and revolving lever, the wheel, cannot be formed of bone and flesh. Man as a motive power is a poor thing. He can only convert three or four thousand calories of energy a day and he does that very inefficiently. But he can make an engine that will handle a hundred thousand times that, twice as efficiently and three times as long. In this way only can he get rid of pain and toil and gain the wealth he wants.

THREE PERIODS OF PROGRESS 13

Gradually then he will substitute for the natural world an artificial world, molded nearer to his heart's desire. Man the Artifex will ultimately master Nature and reign supreme over his own creation until chaos shall come again. In the ancient drama it was *deus ex machina* that came in at the end to solve the problems of the play. It is to the same supernatural agency, the divinity in machinery, that we must look for the salvation of society. It is by means of applied science that the earth can be made habitable and a decent human life made possible. Creative evolution is at last becoming conscious.

II
NITROGEN
PRESERVER AND DESTROYER OF LIFE

In the eyes of the chemist the Great War was essentially a series of explosive reactions resulting in the liberation of nitrogen. Nothing like it has been seen in any previous wars. The first battles were fought with cellulose, mostly in the form of clubs. The next were fought with silica, mostly in the form of flint arrowheads and spear-points. Then came the metals, bronze to begin with and later iron. The nitrogenous era in warfare began when Friar Roger Bacon or Friar Schwartz—whichever it was—ground together in his mortar saltpeter, charcoal and sulfur. The Chinese, to be sure, had invented gunpowder long before, but they—poor innocents—did not know of anything worse to do with it than to make it into fire-crackers. With the introduction of "villainous saltpeter" war ceased to be the vocation of the nobleman and since the nobleman had no other vocation he began to become extinct. A bullet fired from a mile away is no respecter of persons. It is just as likely to kill a knight as a peasant, and a brave man as a coward. You cannot fence with a cannon ball nor overawe it with a plumed hat. The only thing you can do is to hide and shoot back. Now you cannot hide if you send up a column of smoke by day and a pillar of fire by night—the most conspicuous of signals—every time you shoot. So the next step was the invention of a smokeless powder. In this the oxygen necessary for the combustion is already in such close combination with its fuel, the carbon and hydrogen, that no black particles of carbon can get away unburnt. In the old-fashioned gunpowder the oxygen necessary for the combustion of the carbon and sulfur was in a separate package, in the molecule of potassium nitrate, and however finely the mixture was ground, some of the atoms, in the excitement of the explosion, failed to find their proper partners at the moment of dispersal. The new gunpowder besides being smokeless is ashless. There is no black sticky mass of potassium salts left to foul the gun barrel.

The gunpowder period of warfare was actively initiated at the battle of Cressy, in which, as a contemporary historian says, "The English guns made noise like thunder and caused much loss in men and horses." Smokeless powder as invented by Paul Vieille was adopted by the French Government in 1887. This, then, might be called the beginning of the guncotton or nitrocellulose period—or, perhaps in deference to the caveman's club, the second cellulose period of human warfare. Better, doubtless, to call it the "high explosive period," for various other nitro-compounds besides guncotton are being used.

The important thing to note is that all the explosives from gunpowder down contain nitrogen as the essential element. It is customary to call nitrogen "an inert element" because it was hard to get it into combination with other elements. It might, on the other hand, be looked upon as an active element because it acts so energetically in getting out of its compounds. We can dodge the question by saying that nitrogen is a most unreliable and unsociable element. Like Kipling's cat it walks by its wild lone.

It is not so bad as Argon the Lazy and the other celibate gases of that family, where each individual atom goes off by itself and absolutely refuses to unite even temporarily with any other atom. The nitrogen atoms will pair off with each other and stick together, but they are reluctant to associate with other elements and when they do the combination is likely to break up any moment. You all know people like that, good enough when by themselves but sure to break up any club, church or society they get into. Now, the value of nitrogen in warfare is due to the fact that all the atoms desert in a body on the field of battle. Millions of them may be lying packed in a gun cartridge, as quiet as you please, but let a little disturbance start in the neighborhood—say a grain of mercury fulminate flares up—and all the nitrogen atoms get to trembling so violently that they cannot be restrained. The shock spreads rapidly through the whole mass. The hydrogen and carbon atoms catch up the oxygen and in an instant they are off on a stampede, crowding in every direction to find an exit, and getting more heated up all the time. The only movable side is the cannon ball in front, so they all pound against that and give it such a shove that it goes ten miles before it stops. The external bombardment by the cannon ball is, therefore, preceded by an internal bombardment on the cannon ball by the molecules of the hot gases, whose speed is about as great as the speed of the projectile that they propel.

NITROGEN

The active agent in all these explosives is the nitrogen atom in combination with two oxygen atoms, which the chemist calls the "nitro group" and which he represents by NO_2. This group was, as I have said, originally used in the form of saltpeter or potassium nitrate, but since the chemist did not want the potassium part of it—for it fouled his guns—he took the nitro group out of the nitrate by means of sulfuric acid and by the same means hooked it on to some compound of carbon and hydrogen that would burn without leaving any residue, and give nothing but gases. One of the simplest of these hydrocarbon derivatives is glycerin, the same as you use for sunburn. This mixed with nitric and sulfuric acids gives nitroglycerin, an easy thing to make, though I should not advise anybody to try making it unless he has his life insured. But nitroglycerin is uncertain stuff to keep and being a liquid is awkward to handle. So it was mixed with sawdust or porous earth or something else that would soak it up. This molded into sticks is our ordinary dynamite.

If instead of glycerin we take cellulose in the form of wood pulp or cotton and treat this with nitric acid in the presence of sulfuric we get nitrocellulose or guncotton, which is the chief ingredient of smokeless powder.

Now guncotton looks like common cotton. It is too light and loose to pack well into a gun. So it is dissolved with ether and alcohol or acetone to make a plastic mass that can be molded into rods and cut into grains of suitable shape and size to burn at the proper speed.

Here, then, we have a liquid explosive, nitroglycerin,

that has to be soaked up in some porous solid, and a porous solid, guncotton, that has to soak up some liquid. Why not solve both difficulties together by dissolving the guncotton in the nitroglycerin and so get a double explosive? This is a simple idea. Any of us can see the sense of it—once it is suggested to us. But Alfred Nobel, the Swedish chemist, who thought it out first in 1878, made millions out of it. Then, apparently alarmed at the possible consequences of his invention, he bequeathed the fortune he had made by it to found international prizes for medical, chemical and physical discoveries, idealistic literature and the promotion of peace. But his posthumous efforts for the advancement of civilization and the abolition of war did not amount to much and his high explosives were later employed to blow into pieces the doctors, chemists, authors and pacifists he wished to reward.

Nobel's invention, "cordite," is composed of nitroglycerin and nitrocellulose with a little mineral jelly or vaseline. Besides cordite and similar mixtures of nitroglycerin and nitrocellulose there are two other classes of high explosives in common use.

One is made from carbolic acid, which is familiar to us all by its use as a disinfectant. If this is treated with nitric and sulfuric acids we get from it picric acid, a yellow crystalline solid. Every government has its own secret formula for this type of explosive. The British call theirs "lyddite," the French "melinite" and the Japanese "shimose."

The third kind of high explosives uses as its base toluol. This is not so familiar to us as glycerin, cotton or carbolic acid. It is one of the coal tar products, an inflammable liquid, resembling benzene. When treated with nitric acid in the usual way it takes up like the others three nitro groups and so becomes tri-nitro-toluol. Realizing that people could not be expected to use such a mouthful of a word, the chemists have suggested various pretty nicknames, trotyl, tritol, trinol, tolite and trilit, but the public, with the wilfulness it always shows in the matter of names, persists in calling it TNT, as though it were an author like G. B. S., or G. K. C., or F. P. A. TNT is the latest of these high explosives and in some ways the best of them. Picric acid has the bad habit of attacking the metals with which it rests in contact forming sensitive picrates that are easily set off, but TNT is inert toward metals and keeps well. TNT melts far below the boiling point of water so can be readily liquefied and poured into shells. It is insensitive to ordinary shocks. A rifle bullet can be fired through a case of it without setting it off, and if lighted with a match it burns quietly. The amazing thing about these modern explosives, the organic nitrates, is the way they will stand banging about and burning, yet the terrific violence with which they blow up when shaken by an explosive wave of a particular velocity like that of a fulminating cap. Like picric acid, TNT stains the skin yellow and causes soreness and sometimes serious cases of poisoning among the employees, mostly girls, in the munition factories. On the other hand, the girls working with cordite get to using it as chewing gum; a harmful habit, not because of any danger of being blown up by it, but because nitroglycerin is a heart stimulant and they do not need that.

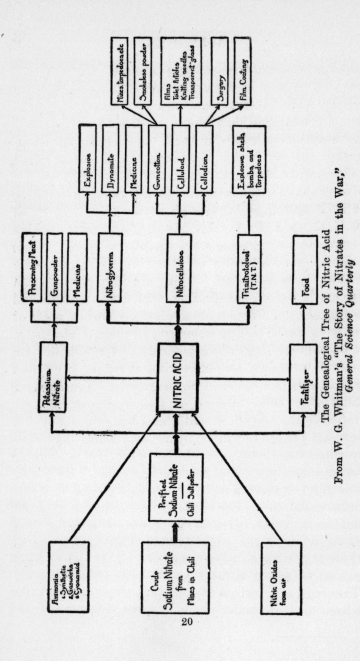

The Genealogical Tree of Nitric Acid

From W. G. Whitman's "The Story of Nitrates in the War," *General Science Quarterly*

TNT is by no means smokeless. The German shells that exploded with a cloud of black smoke and which British soldiers called "Black Marias," "coal-boxes" or "Jack Johnsons" were loaded with it. But it is an advantage to have a shell show where it strikes, although a disadvantage to have it show where it starts.
It is these high explosives that have revolutionized warfare. As soon as the first German shell packed with these new nitrates burst inside the Gruson cupola at Liège and tore out its steel and concrete by the roots the world knew that the day of the fixed fortress was gone. The armies deserted their expensively prepared fortifications and took to the trenches. The British troops in France found their weapons futile and sent across the Channel the cry of "Send us high explosives or we perish!" The home Government was slow to heed the appeal, but no progress was made against the Germans until the Allies had the means to blast them out of their entrenchments by shells loaded with five hundred pounds of TNT.
All these explosives are made from nitric acid and this used to be made from nitrates such as potassium nitrate or saltpeter. But nitrates are rarely found in large quantities. Napoleon and Lee had a hard time to scrape up enough saltpeter from the compost heaps, cellars and caves for their gunpowder, and they did not use as much nitrogen in a whole campaign as was freed in a few days' cannonading on the Somme. Now there is one place in the world—and so far as we know one only—where nitrates are to be found abundantly. This is in a desert on the western slope of the Andes where ancient guano deposits have decomposed and

there was not enough rain to wash away their salts. Here is a bed two miles wide, two hundred miles long and five feet deep yielding some twenty to fifty per cent. of sodium nitrate. The deposit originally belonged to Peru, but Chile fought her for it and got it in 1881. Here all countries came to get their nitrates for agriculture and powder making. Germany was the largest customer and imported 750,000 tons of Chilean nitrate in 1913, besides using 100,000 tons of other nitrogen salts. By this means her old, wornout fields were made to yield greater harvests than our fresh land. Germany and England were like two duelists buying powder at the same shop. The Chilean Government, pocketing an export duty that aggregated half a billion dollars, permitted the saltpeter to be shoveled impartially into British and German ships, and so two nitrogen atoms, torn from their Pacific home and parted, like Evangeline and Gabriel, by transportation oversea, may have found themselves flung into each other's arms from the mouths of opposing howitzers in the air of Flanders. Goethe could write a romance on such a theme.

Now the moment war broke out this source of supply was shut off to both parties, for they blockaded each other. The British fleet closed up the German ports while the German cruisers in the Pacific took up a position off the coast of Chile in order to intercept the ships carrying nitrates to England and France. The Panama Canal, designed to afford relief in such an emergency, caved in most inopportunely. The British sent a fleet to the Pacific to clear the nitrate route, but it was outranged and defeated on November 1, 1914. Then a

NITROGEN

stronger British fleet was sent out and smashed the Germans off the Falkland Islands on December 8. But for seven weeks the nitrate route had been closed while the chemical reactions on the Marne and Yser were decomposing nitrogen-compounds at an unheard of rate.

England was now free to get nitrates for her munition factories, but Germany was still bottled up. She had stored up Chilean nitrates in anticipation of the war and as soon as it was seen to be coming she bought all she could get in Europe. But this supply was altogether inadequate and the war would have come to an end in the first winter if German chemists had not provided for such a contingency in advance by working out methods of getting nitrogen from the air. Long ago it was said that the British ruled the sea and the French the land so that left nothing to the German but the air. The Germans seem to have taken this jibe seriously and to have set themselves to make the most of the aerial realm in order to challenge the British and French in the fields they had appropriated. They had succeeded so far that the Kaiser when he declared war might well have considered himself the Prince of the Power of the Air. He had a fleet of Zeppelins and he had means for the fixation of nitrogen such as no other nation possessed. The Zeppelins burst like wind bags, but the nitrogen plants worked and made Germany independent of Chile not only during the war, but in the time of peace.

Germany during the war used 200,000 tons of nitric acid a year in explosives, yet her supply of nitrogen is exhaustless.

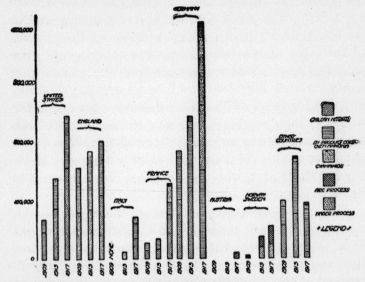

World production and consumption of fixed inorganic nitrogen expressed in tons nitrogen

From *The Journal of Industrial and Engineering Chemistry*, March, 1919.

Nitrogen is free as air. That is the trouble; it is too free. It is fixed nitrogen that we want and that we are willing to pay for; nitrogen in combination with some other elements in the form of food or fertilizer so we can make use of it as we set it free. Fixed nitrogen in its cheapest form, Chile saltpeter, rose to $250 during the war. Free nitrogen costs nothing and is good for nothing. If a land-owner has a right to an expanding pyramid of air above him to the limits of the atmosphere—as, I believe, the courts have decided in the eaves-dropping cases—then for every square foot of his ground he owns as much nitrogen as he could buy for $2500. The air is four-fifths free nitrogen and if we could absorb it in our lungs as we do the oxygen of the other fifth a few minutes breathing would give us a full meal. But we let this free nitrogen all out again through our noses and then go and pay 35 cents a pound for steak or 60 cents a dozen for eggs in order to get enough combined nitrogen to live on. Though man is immersed in an ocean of nitrogen, yet he cannot make use of it. He is like Coleridge's "Ancient Mariner" with "water, water, everywhere, nor any drop to drink."

Nitrogen is, as Hood said not so truly about gold, "hard to get and hard to hold." The bacteria that form the nodules on the roots of peas and beans have the power that man has not of utilizing free nitrogen. Instead of this quiet inconspicuous process man has to call upon the lightning when he wants to fix nitrogen. The air contains the oxygen and nitrogen which it is desired to combine to form nitrates but the atoms are paired, like to like. Passing an electric spark through the air breaks up some of these pairs and in the confusion of the shock the lonely atoms seize on their nearest neighbor and so may get partners of the other sort. I have seen this same thing happen in a square dance where somebody made a blunder. It is easy to understand the reaction if we represent the atoms of oxygen and nitrogen by the initials of their names in this fashion:

$$NN + OO \rightarrow NO + NO$$
$$\text{nitrogen} \quad \text{oxygen} \quad \text{nitric oxide}$$

The → represents Jove's thunderbolt, a stroke of artificial lightning. We see on the left the molecules of oxygen and nitrogen, before taking the electric treatment, as separate elemental pairs, and then to the right of the arrow we find them as compound molecules of nitric oxide. This takes up another atom of oxygen from the air and becomes NOO, or using a subscript figure to indicate the number of atoms and so avoid repeating the letter, NO_2 which is the familiar nitro group of nitric acid ($HO-NO_2$) and of its salts, the nitrates, and of its organic compounds, the high explosives. The NO_2 is a brown and evil-smelling gas which when dissolved in water (HOH) and further oxidized is completely converted into nitric acid.

The apparatus which effects this transformation is essentially a gigantic arc light in a chimney through which a current of hot air is blown. The more thoroughly the air comes under the action of the electric arc the more molecules of nitrogen and oxygen will be broken up and rearranged, but on the other hand if the mixture of gases remains in the path of the discharge the NO molecules are also broken up and go back into their original form of NN and OO. So the object is to spread out the electric arc as widely as possible and then run the air through it rapidly. In the Schönherr process the electric arc is a spiral flame twenty-three feet long through which the air streams with a vortex motion. In the Birkeland-Eyde furnace there is a series of semi-circular arcs spread out by the repellent force of a powerful electric magnet in a flaming disc seven feet in diameter with a temperature of 6300° F. In the Pauling furnace the electrodes between which the current strikes are two cast iron tubes curving upward and outward like the horns of a Texas steer and cooled by a stream of water passing through them. These electric furnaces produce two or three ounces of nitric acid for each kilowatt-hour of current consumed. Whether they can compete with the natural nitrates and the products of other processes depends upon how cheaply they can get their electricity. Before the war there were several large installations in Norway and elsewhere where abundant water power was available and now the Norwegians are using half a million horse power continuously in the fixation of nitrogen and the rest of the world as much again. The Germans had invested largely in these foreign oxidation plants, but shortly before the war they had sold out and turned their attention to other processes not requiring so much electrical energy, for their country is poorly provided with water power. The Haber process, that they made most of, is based upon as simple a reaction as that we have been considering, for it consists in uniting two elemental gases to make a compound, but the elements in this case are not nitrogen and oxygen, but nitrogen and hydrogen. This gives ammonia instead of nitric acid, but ammonia is useful for its own purposes and it can be converted into nitric acid if this is desired. The reaction is:

$$\underset{\text{nitrogen}}{\text{NN}} + \underset{\text{hydrogen}}{\text{HH} + \text{HH} + \text{HH}} \rightarrow \underset{\text{ammonia}}{\text{NHHH} + \text{NHHH}}$$

The animals go in two by two, but they come out four by four. Four molecules of the mixed elements are turned into two molecules and so the gas shrinks to half its volume. At the same time it acquires an odor—familiar to us when we are curing a cold—that neither of the original gases had. The agent that effects the transformation in this case is not the electric spark—for this would tend to work the reaction backwards—but uranium, a rare metal, which has the peculiar property of helping along a reaction while seeming to take no part in it. Such a substance is called a catalyst. The action of a catalyst is rather mysterious and whenever we have a mystery we need an analogy. We may, then, compare the catalyst to what is known as "a good mixer" in society. You know the sort of man I mean. He may not be brilliant or especially talkative, but somehow there is always "something doing" at a picnic or house-party when he is along. The tactful hostess, the salon leader, is a social catalyst. The trouble with catalysts, either human or metallic, is that they are rare and that sometimes they get sulky and won't work if the ingredients they are supposed to mix are unsuitable.

But the uranium, osmium, platinum or whatever metal is used as a catalyzing agent is expensive and although it is not used up it is easily "poisoned," as the chemists say, by impurities in the gases. The nitrogen and the hydrogen for the Haber process must then be prepared and purified before trying to combine them into ammonia. The nitrogen is obtained by liquefying air by cold and pressure and then boiling off the nitrogen at $-194°$ C. The oxygen left is useful for other purposes. The hydrogen needed is extracted by a similar process of fractional distillation from "water-gas," the blue-flame burning gas used for heating.

Then the nitrogen and hydrogen, mixed in the proportion of one to three, as shown in the reaction given above, are compressed to two hundred atmospheres, heated to 1300° F. and passed over the finely divided uranium. The stream of gas that comes out contains about four per cent. of ammonia, which is condensed to a liquid by cooling and the uncombined hydrogen and nitrogen passed again through the apparatus.

The ammonia can be employed in refrigeration and other ways but if it is desired to get the nitrogen into the form of nitric acid it has to be oxidized by the so-called Ostwald process. This is the reaction:

$$\underset{\text{ammonia}}{NH_3} + \underset{\text{oxygen}}{4O} \rightarrow \underset{\text{nitric acid}}{HNO_3} + \underset{\text{water}}{H_2O}$$

The catalyst used to effect this combination is the metal platinum in the form of fine wire gauze, since the action takes place only on the surface. The ammonia gas is mixed with air which supplies the oxygen and the heated mixture run through the platinum gauze at the rate of several yards a second. Although the gases come in contact with the platinum only a five-hundredth part of a second yet eighty-five per cent. is converted into nitric acid.

The Haber process for the making of ammonia by direct synthesis from its constituent elements and the supplemental Ostwald process for the conversion of the ammonia into nitric acid were the salvation of Germany. As soon as the Germans saw that their dash toward Paris had been stopped at the Marne they knew that they were in for a long war and at once made plans for a supply of fixed nitrogen. The chief German dye factories, the Badische Anilin and Soda-Fabrik, promptly put $100,000,000 into enlarging its plant and raised its production of ammonium sulfate from 30,000 to 300,000 tons. One German electrical firm with aid from the city of Berlin contracted to provide 66,000,000 pounds of fixed nitrogen a year at a cost of three cents a pound for the next twenty-five years. The 750,000 tons of Chilean nitrate imported annually by Germany contained about 116,000 tons of the essential element nitrogen. The fourteen large plants erected during the war can fix in the form of nitrates 500,000 tons of nitrogen a year, which is more than twice the amount needed for internal consumption. So Germany is now not only independent of the outside world but will have a surplus of nitrogen products which could be sold even in America at about half what the farmer has been paying for South American saltpeter.

Besides the Haber or direct process there are other methods of making ammonia which are, at least outside of Germany, of more importance. Most prominent of these is the cyanamid process. This requires electrical power since it starts with a product of the electrical furnace, calcium carbide, familiar to us all as a source of acetylene gas.

If a stream of nitrogen is passed over hot calcium carbide it is taken up by the carbide according to the following equation:

$$\underset{\text{calcium carbide}}{CaC_2} + \underset{\text{nitrogen}}{N_2} \rightarrow \underset{\text{calcium cyanamid}}{CaCN_2} + \underset{\text{carbon}}{C}$$

Calcium cyanamid was discovered in 1895 by Caro and Franke when they were trying to work out a new

NITROGEN

process for making cyanide to use in extracting gold. It looks like stone and, under the name of lime-nitrogen, or Kalkstickstoff, or nitrolim, is sold as a fertilizer. If it is desired to get ammonia, it is treated with superheated steam. The reaction produces heat and pressure, so it is necessary to carry it on in stout autoclaves or enclosed kettles. The cyanamid is completely and quickly converted into pure ammonia and calcium carbonate, which is the same as the limestone from which carbide was made. The reaction is:

$$\underset{\text{calcium cyanamid}}{CaCN_2} + \underset{\text{water}}{3H_2O} \rightarrow \underset{\text{calcium carbonate}}{CaCO_3} + \underset{\text{ammonia}}{2NH_3}$$

Another electrical furnace method, the Serpek process, uses aluminum instead of calcium for the fixation of nitrogen. Bauxite, or impure aluminum oxide, the ordinary mineral used in the manufacture of metallic aluminum, is mixed with coal and heated in a revolving electrical furnace through which nitrogen is passing. The equation is:

$$\underset{\text{aluminum oxide}}{Al_2O_3} + \underset{\text{carbon}}{3C} + \underset{\text{nitrogen}}{N_2} \rightarrow \underset{\text{aluminum nitride}}{2AlN} + \underset{\text{carbon monoxide}}{3CO}$$

Then the aluminum nitride is treated with steam under pressure, which produces ammonia and gives back the original aluminum oxide, but in a purer form than the mineral from which was made

$$\underset{\text{aluminum nitride}}{2AlN} + \underset{\text{water}}{3H_2O} \rightarrow \underset{\text{ammonia}}{2NH_3} + \underset{\text{aluminum oxide}}{Al_2O_3}$$

The Serpek process is employed to some extent in France in connection with the aluminum industry. These are the principal processes for the fixation of nitrogen now in use, but they by no means exhaust the possibilities. For instance, Professor John C. Bucher, of Brown University, created a sensation in 1917 by announcing a new process which he had worked out with admirable completeness and which has some very attractive features. It needs no electric power or high pressure retorts or liquid air apparatus. He simply fills a twenty-foot tube with briquets made out of soda ash, iron and coke and passes producer gas through the heated tube. Producer gas contains nitrogen since it is made by passing air over hot coal. The reaction is:

$$2Na_2CO_3 + 4C + N_2 = 2NaCN + 3CO$$
sodium carbonate + carbon + nitrogen = sodium cyanide + carbon monoxide

The iron here acts as the catalyst and converts two harmless substances, sodium carbonate, which is common washing soda, and carbon, into two of the most deadly compounds known to man, cyanide and carbon monoxide, which is what kills you when you blow out the gas. Sodium cyanide is a salt of hydrocyanic acid, which for some curious reason is called "Prussic acid." It is so violent a poison that, as the freshman said in a chemistry recitation, "a single drop of it placed on the tongue of a dog will kill a man."

But sodium cyanide is not only useful in itself, for the extraction of gold and cleaning of silver, but can be converted into ammonia, and a variety of other compounds such as urea and oxamid, which are good fertilizers; sodium ferrocyanide, that makes Prussian blue; and oxalic acid used in dyeing. Professor Bucher claimed that his furnace could be set up in a day at a cost of less than $100 and could turn out 150 pounds of

NITROGEN

sodium cyanide in twenty-four hours. This process was placed freely at the disposal of the United States Government for the war and a 10-ton plant was built at Saltville, Va., by the Ordnance Department. But the armistice put a stop to its operations and left the future of the process undetermined.

We might have expected that the fixation of nitrogen by passing an electrical spark through hot air would have been an American invention, since it was Franklin who snatched the lightning from the heavens as well as the scepter from the tyrant and since our output of hot air is unequaled by any other nation. But little attention was paid to the nitrogen problem until 1916 when it became evident that we should soon be drawn into a war "with a first class power." On June 3, 1916, Congress placed $20,000,000 at the disposal of the president for investigation of "the best, cheapest and most available means for the production of nitrate and other products for munitions of war and useful in the manufacture of fertilizers and other useful products by water power or any other power." But by the time war was declared on April 6, 1917, no definite program had been approved and by the time the armistice was signed on November 11, 1918, no plants were in active operation. But five plants had been started and two of them were nearly ready to begin work when they were closed by the ending of the war. United States Nitrate Plant No. 1 was located at Sheffield, Alabama, and was designed for the production of ammonia by "direct action" from nitrogen and hydrogen according to the plans of the American Chemical Company. Its capacity was calculated at 60,000 pounds of anhydrous ammonia a day, half of which was to be oxidized to nitric acid. Plant No. 2 was erected at Muscle Shoals, Alabama, to use the process of the American Cyanamid Company. This was contracted to produce 110,000 tons of ammonium nitrate a year and later two other cyanamid plants of half that capacity were started at Toledo and Ancor, Ohio.

At Muscle Shoals a mushroom city of 20,000 sprang up on an Alabama cotton field in six months. The raw material, air, was as abundant there as anywhere and the power, water, could be obtained from the Government hydro-electric plant on the Tennessee River, but this was not available during the war, so steam was employed instead. The heat of the coal was used to cool the air down to the liquefying point. The principle of this process is simple. Everybody knows that heat expands and cold contracts, but not everybody has realized the converse of this rule, that expansion cools and compression heats. If air is forced into smaller space, as in a tire pump, it heats up and if allowed to expand to ordinary pressure it cools off again. But if the air while compressed is cooled and then allowed to expand it must get still colder and the process can go on till it becomes cold enough to congeal. That is, by expanding a great deal of air, a little of it can be reduced to the liquefying point. At Muscle Shoals the plant for liquefying air, in order to get the nitrogen out of it, consisted of two dozen towers each capable of producing 1765 cubic feet of pure nitrogen per hour. The air was drawn in through two pipes, a yard across, and passed through scrubbing towers to remove impurities. The air was then compressed to 600 pounds per square

NITROGEN

inch. Nine tenths of the air was permitted to expand to 50 pounds and this expansion cooled down the other tenth, still under high pressure, to the liquefying point. Rectifying towers 24 feet high were stacked with trays of liquid air from which the nitrogen was continually bubbling off since its boiling point is twelve degrees centigrade lower than that of oxygen. Pure nitrogen gas collected at the top of the tower and the residual liquid air, now about half oxygen, was allowed to escape at the bottom.

The nitrogen was then run through pipes into the lime-nitrogen ovens. There were 1536 of these about four feet square and each holding 1600 pounds of pulverized calcium carbide. This is at first heated by an electrical current to start the reaction which afterwards produces enough heat to keep it going. As the stream of nitrogen gas passes over the finely divided carbide it is absorbed to form calcium cyanamid as described on a previous page. This product is cooled, powdered and wet to destroy any quicklime or carbide left unchanged. Then it is charged into autoclaves and steam at high temperature and pressure is admitted. The steam acting on the cyanamid sets free ammonia gas which is carried to towers down which cold water is sprayed, giving the ammonia water, familiar to the kitchen and the bathroom.

But since nitric acid rather than ammonia was needed for munitions, the oxygen of the air had to be called into play. This process, as already explained, is carried on by aid of a catalyzer, in this case platinum wire. At Muscle Shoals there were 696 of these catalyzer boxes. The ammonia gas, mixed with air to provide the necessary oxygen, was admitted at the top and passed down through a sheet of platinum gauze of 80 mesh to the inch, heated to incandescence by electricity. In contact with this the ammonia is converted into gaseous oxides of nitrogen (the familiar red fumes of the laboratory) which, carried off in pipes, cooled and dissolved in water, form nitric acid.

But since none of the national plants could be got into action during the war, the United States was compelled to draw upon South America for its supply. The imports of Chilean saltpeter rose from half a million tons in 1914 to a million and a half in 1917. After peace was made the Department of War turned over to the Department of Agriculture its surplus of saltpeter, 150,000 tons, and it was sold to American farmers at cost, $81 a ton.

For nitrogen plays a double rôle in human economy. It appears like Brahma in two aspects, Vishnu the Preserver and Siva the Destroyer. Here I have been considering nitrogen in its maleficent aspect, its use in war. We now turn to its beneficent aspect, its use in peace.

III

FEEDING THE SOIL

The Great War not only starved people: it starved the land. Enough nitrogen was thrown away in some indecisive battle on the Aisne to save India from a famine. The population of Europe as a whole has not been lessened by the war, but the soil has been robbed of its power to support the population. A plant requires certain chemical elements for its growth and all of these must be within reach of its rootlets, for it will accept no substitutes. A wheat stalk in France before the war had placed at its feet nitrates from Chile, phosphates from Florida and potash from Germany. All these were shut off by the firing line and the shortage of shipping.

Out of the eighty elements only thirteen are necessary for crops. Four of these are gases: hydrogen, oxygen, nitrogen and chlorine. Five are metals: potassium, magnesium, calcium, iron and sodium. Four are non-metallic solids: carbon, sulfur, phosphorus and silicon. Three of these, hydrogen, oxygen and carbon, making up the bulk of the plant, are obtainable *ad libitum* from the air and water. The other ten in the form of salts are dissolved in the water that is sucked up from the soil. The quantity needed by the plant is so small and the quantity contained in the soil is so great that ordinarily we need not bother about the supply except in case of three of them. They are nitrogen, potassium and phosphorus. These would be useless or fatal to plant life in the elemental form, but fixed in neutral salt they are essential plant foods. A ton of wheat takes away from the soil about 47 pounds of nitrogen, 18 pounds of phosphoric acid and 12 pounds of potash. If then the farmer does not restore this much to his field every year he is drawing upon his capital and this must lead to bankruptcy in the long run.

So much is easy to see, but actually the question is extremely complicated. When the German chemist, Justus von Liebig, pointed out in 1840 the possibility of maintaining soil fertility by the application of chemicals it seemed at first as though the question were practically solved. Chemists assumed that all they had to do was to analyze the soil and analyze the crop and from this figure out, as easily as balancing a bank book, just how much of each ingredient would have to be restored to the soil every year. But somehow it did not work out that way and the practical agriculturist, finding that the formulas did not fit his farm, sneered at the professors and whenever they cited Liebig to him he irreverently transposed the syllables of the name. The chemist when he went deeper into the subject saw that he had to deal with the colloids, damp, unpleasant, gummy bodies that he had hitherto fought shy of because they would not crystallize or filter. So the chemist called to his aid the physicist on the one hand and the biologist on the other and then they both had their hands full. The physicist found that he had to deal with a polyvariant system of solids, liquids and gases

FEEDING THE SOIL

mutually miscible in phases too numerous to be handled by Gibbs's Rule. The biologist found that he had to deal with the invisible flora and fauna of a new world.

Plants obey the injunction of Tennyson and rise on the stepping stones of their dead selves to higher things. Each successive generation lives on what is left of the last in the soil plus what it adds from the air and sunshine. As soon as a leaf or tree trunk falls to the ground it is taken in charge by a wrecking crew composed of a myriad of microscopic organisms who proceed to break it up into its component parts so these can be used for building a new edifice. The process is called "rotting" and the product, the black, gummy stuff of a fertile soil, is called "humus." The plants, that is, the higher plants, are not able to live on their own proteids as the animals are. But there are lower plants, certain kinds of bacteria, that can break up the big complicated proteid molecules into their component parts and reduce the nitrogen in them to ammonia or ammonia-like compounds. Having done this they stop and turn over the job to another set of bacteria to be carried through the next step. For you must know that soil society is as complex and specialized as that above ground and the tiniest bacterium would die rather than violate the union rules. The second set of bacteria change the ammonia over to nitrites and then a third set, the Amalgamated Union of Nitrate Workers, steps in and completes the process of oxidation with an efficiency that Ostwald might envy, for ninety-six per cent. of the ammonia of the soil is converted into nitrates. But if the conditions are not just right, if the food is insufficient or unwholesome or if the air that circulates through the soil is contaminated with poison gases, the bacteria go on a strike. The farmer, not seeing the thing from the standpoint of the bacteria, says the soil is "sick" and he proceeds to doctor it according to his own notion of what ails it. First perhaps he tries running in strike breakers. He goes to one of the firms that makes a business of supplying nitrogen-fixing bacteria from the scabs or nodules of the clover roots and scatters these colonies over the field. But if the living conditions remain bad the newcomers will soon quit work too and the farmer loses his money. If he is wise, then, he will remedy the conditions, putting a better ventilation system in his soil perhaps or neutralizing the sourness by means of lime or killing off the ameboid banditti that prey upon the peaceful bacteria engaged in the nitrogen industry. It is not an easy job that the farmer has in keeping billions of billions of subterranean servants contented and working together, but if he does not succeed at this he wastes his seed and labor.

The layman regards the soil as a platform or anchoring place on which to set plants. He measures its value by its superficial area without considering its contents, which is as absurd as to estimate a man's wealth by the size of his safe. The difference in point of view is well illustrated by the old story of the city chap who was showing his farmer uncle the sights of New York. When he took him to Central Park he tried to astonish him by saying "This land is worth $500,000 an acre." The old farmer dug his toe into the ground, kicked out a clod, broke it open, looked at it, spit on it

FEEDING THE SOIL

and squeezed it in his hand and then said, "Don't you believe it; 'tain't worth ten dollars an acre. Mighty poor soil I call it." Both were right.

The modern agriculturist realizes that the soil is a laboratory for the production of plant food and he ordinarily takes more pains to provide a balanced ration for it than he does for his family. Of course the necessity of feeding the soil has been known ever since man began to settle down and the ancient methods of maintaining its fertility, though discovered accidentally and followed blindly, were sound and efficacious. Virgil, who like Liberty Hyde Bailey was fond of publishing agricultural bulletins in poetry, wrote two thousand years ago:

> But sweet vicissitudes of rest and toil
> Make easy labor and renew the soil
> Yet sprinkle sordid ashes all around
> And load with fatt'ning dung thy fallow soil.

The ashes supplied the potash and the dung the nitrate and phosphate. Long before the discovery of the nitrogen-fixing bacteria, the custom prevailed of sowing pea-like plants every third year and then plowing them under to enrich the soil. But such local supplies were always inadequate and as soon as deposits of fertilizers were discovered anywhere in the world they were drawn upon. The richest of these was the Chincha Islands off the coast of Peru, where millions of penguins and pelicans had lived in a most untidy manner for untold centuries. The guano composed of the excrement of the birds mixed with the remains of dead birds and the fishes they fed upon was piled up to a depth of 120 feet. From this Isle of Penguins—which is not that described by Anatole France—a billion dollars' worth of guano was taken and the deposit was soon exhausted.

Then the attention of the world was directed to the mainland of Peru and Chile, where similar guano deposits had been accumulated and, not being washed away on account of the lack of rain, had been deposited as sodium nitrate, or "saltpeter." These beds were discovered by a German, Taddeo Haenke, in 1809, but it was not until the last quarter of the century that the nitrates came into common use as a fertilizer. Since then more than 53,000,000 tons have been taken out of these beds and the exportation has risen to a rate of 2,500,000 to 3,000,000 tons a year. How much longer they will last is a matter of opinion and opinion is largely influenced by whether you have your money invested in Chilean nitrate stock or in one of the new synthetic processes for making nitrates. The United States Department of Agriculture says the nitrate beds will be exhausted in a few years. On the other hand the Chilean Inspector General of Nitrate Deposits in his latest official report says that they will last for two hundred years at the present rate and that then there are incalculable areas of low grade deposits, containing less than eleven per cent., to be drawn upon.

Anyhow, the South American beds cannot long supply the world's need of nitrates and we shall some time be starving unless creative chemistry comes to the rescue. In 1898 Sir William Crookes—the discoverer of the "Crookes tubes," the radiometer and radiant matter—startled the British Association for the Advance-

FEEDING THE SOIL

ment of Science by declaring that the world was nearing the limit of wheat production and that by 1931 the bread-eaters, the Caucasians, would have to turn to other grains or restrict their population while the rice and millet eaters of Asia would continue to increase. Sir William was laughed at then as a sensationalist. He was, but his sensations were apt to prove true and it is already evident that he was too near right for comfort. Before we were half way to the date he set we had two wheatless days a week, though that was because we persisted in shooting nitrates into the air. The area producing wheat was by decades:[1]

THE WHEAT FIELDS OF THE WORLD

	Acres
1881–90	192,000,000
1890–1900	211,000,000
1900–10	242,000,000
Probable limit	300,000,000

If 300,000,000 acres can be brought under cultivation for wheat and the average yield raised to twenty bushels to the acre, that will give enough to feed a billion people if they eat six bushels a year as do the English. Whether this maximum is correct or not there is evidently some limit to the area which has suitable soil and climate for growing wheat, so we are ultimately thrown back upon Crookes's solution of the problem; that is, we must increase the yield per acre and this can only be done by the use of fertilizers and especially by the fixation of atmospheric nitrogen. Crookes esti-

[1] I am quoting mostly Unstead's figures from the *Geographical Journal* of 1913. See also Dickson's "The Distribution of Mankind," in Smithsonian Report, 1913.

mated the average yield of wheat at 12.7 bushels to the acre, which is more than it is in the new lands of the United States, Australia and Russia, but less than in Europe, where the soil is well fed. What can be done to increase the yield may be seen from these figures:

GAIN IN THE YIELD OF WHEAT IN BUSHELS PER ACRE

	1889–90	1913
Germany	19	35
Belgium	30	35
France	17	20
United Kingdom	28	32
United States	12	15

The greatest gain was made in Germany and we see a reason for it in the fact that the German importation of Chilean saltpeter was 55,000 tons in 1880 and 747,000 tons in 1913. In potatoes, too, Germany gets twice as big a crop from the same ground as we do, 223 bushels per acre instead of our 113 bushels. But the United States uses on the average only 28 pounds of fertilizer per acre, while Europe uses 200.

It is clear that we cannot rely upon Chile, but make nitrates for ourselves as Germany had to in war time. In the first chapter we considered the new methods of fixing the free nitrogen from the air. But the fixation of nitrogen is a new business in this country and our chief reliance so far has been the coke ovens. When coal is heated in retorts or ovens for making coke or gas a lot of ammonia comes off with the other products of decomposition and is caught in the sulfuric acid used to wash the gas as ammonium sulfate. Our American coke-makers have been in the habit of letting

FEEDING THE SOIL

this escape into the air and consequently we have been losing some 700,000 tons of ammonium salts every year, enough to keep our land rich and give us all the explosives we should need. But now they are reforming and putting in ovens that save the by-products such as

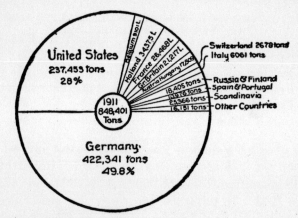

Courtesy of *Scientific American*.
Consumption of potash for agricultural purposes in different countries

ammonia and coal tar, so in 1916 we got from this source 325,000 tons a year.

Germany had a natural monopoly of potash as Chile had a natural monopoly of nitrates. The agriculture of Europe and America has been virtually dependent upon these two sources of plant foods. Now when the world was cleft in twain by the shock of August, 1914, the Allied Powers had the nitrates and the Central Powers had the potash. If Germany had not had up her sleeve a new process for making nitrates she could not long have carried on a war and doubtless would not have ventured upon it. But the outside world had no such substitute for the German potash salts and has not yet discovered one. Consequently the price of potash in the United States jumped from $40 to $400 and the cost of food went up with it. Even under the stimulus of prices ten times the normal and with chem-

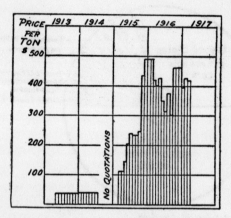

What happened to potash when the war broke out. This diagram from the *Journal of Industrial and Engineering Chemistry* of July, 1917, shows how the supply of potassium muriate from Germany was shut off in 1914 and how its price rose.

ists searching furnace crannies and bad lands the United States was able to scrape up less than 10,000 tons of potash in 1916, and this was barely enough to satisfy our needs for two weeks!

Yet potash compounds are as cheap as dirt. Pick up a handful of gravel and you will be able to find much of it feldspar or other mineral containing some ten per cent. of potash. Unfortunately it is in combination with silica, which is harder to break up than a trust.

But "constant washing wears away stones" and the

potash that the metallurgist finds too hard to extract in his hottest furnace is washed out in the course of time through the dropping of the gentle rain from heaven. "All rivers run to the sea" and so the sea gets salt, all sorts of salts, principally sodium chloride (our table salt) and next magnesium, calcium and potassium chlorides or sulfates in this order of abundance. But if we evaporate sea-water down to dryness all these are left in a mix together and it is hard to sort them out. Only patient Nature has time for it and she only did on a large scale in one place, that is at Stassfurt, Germany. It seems that in the days when northwestern Prussia was undetermined whether it should be sea or land it was flooded annually by seawater. As this slowly evaporated the dissolved salts crystallized out at the critical points, leaving beds of various combinations. Each year there would be deposited three to five inches of salts with a thin layer of calcium sulfate or gypsum on top. Counting these annual layers, like the rings on a stump, we find that the Stassfurt beds were ten thousand years in the making. They were first worked for their salt, common salt, alone, but in 1837 the Prussian Government began prospecting for new and deeper deposits and found, not the clean rock salt that they wanted, but bittern, largely magnesium sulfate or Epsom salt, which is not at all nice for table use. This stuff was first thrown away until it was realized that it was much more valuable for the potash it contains than was the rock salt they were after. Then the Germans began to purify the Stassfurt salts and market them throughout the world. They contain from fifteen to twenty-five per cent. of magnesium chloride mixed with magnesium chloride in "carnallite," with magnesium sulfate in "kainite" and sodium chloride in "sylvinite." More than thirty thousand miners and workmen are employed in the Stassfurt works. There are some seventy distinct establishments engaged in the business, but they are in combination. In fact they are compelled to be, for the German Government is as anxious to promote trusts as the American Government is to prevent them. Once the Stassfurt firms had a falling out and began a cut-throat competition. But the German Government objects to its people cutting each other's throats. American dealers were getting unheard of bargains when the German Government stepped in and compelled the competing corporations to recombine under threat of putting on an export duty that would eat up their profits.

The advantages of such business coöperation are specially shown in opening up a new market for an unknown product as in the case of the introduction of the Stassfurt salts into American agriculture. The farmer in any country is apt to be set in his ways and when it comes to inducing him to spend his hard-earned money for chemicals that he never heard of and could not pronounce he—quite rightly—has to be shown. Well, he was shown. It was, if I remember right, early in the nineties that the German Kali Syndikat began operations in America and the United States Government became its chief advertising agent. In every state there was an agricultural experiment station and these were provided liberally with illustrated literature on Stassfurt salts with colored wall charts and sets of samples and free sacks of salts for field experiments.

The station men, finding that they could rely upon the scientific accuracy of the information supplied by Kali and that the experiments worked out well, became enthusiastic advocates of potash fertilizers. The station bulletins—which Uncle Sam was kind enough to carry free to all the farmers of the state—sometimes were worded so like the Kali Company advertising that the company might have raised a complaint of plagiarizing, but they never did. The Chilean nitrates, which are under British control, were later introduced by similar methods through the agency of the state agricultural experiment stations.

As a result of all this missionary work, which cost the Kali Company $50,000 a year, the attention of a large proportion of American farmers was turned toward intensive farming and they began to realize the necessity of feeding the soil that was feeding them. They grew dependent upon these two foreign and widely separated sources of supply. In the year before the war the United States imported a million tons of Stassfurt salts, for which the farmers paid more than $20,000,000. Then a declaration of American independence—the German embargo of 1915—cut us off from Stassfurt and for five years we had to rely upon our own resources. We have seen how Germany—shut off from Chile—solved the nitrogen problem for her fields and munition plants. It was not so easy for us—shut off from Germany—to solve the potash problem.

There is no more lack of potash in the rocks than there is of nitrogen in the air, but the nitrogen is free and has only to be caught and combined, while the potash is shut up in a granite prison from which it is hard to get it free. It is not the percentage in the soil but the percentage in the soil water that counts. A farmer with his potash locked up in silicates is like the merchant who has left the key of his safe at home in his other trousers. He may be solvent, but he cannot meet a sight draft. It is only solvent potash that passes current.

In the days of our grandfathers we had not only national independence but household independence. Every homestead had its own potash plant and soap factory. The frugal housewife dumped the maple wood ashes of the fireplace into a hollow log set up on end in the backyard. Water poured over the ashes leached out the lye, which drained into a bucket beneath. This gave her a solution of pearl ash or potassium carbonate whose concentration she tested with an egg as a hydrometer. In the meantime she had been saving up all the waste grease from the frying pan and pork rinds from the plate and by trying out these she got her soap fat. Then on a day set apart for this disagreeable process in chemical technology she boiled the fat and the lye together and got "soft soap," or as the chemist would call it, potassium stearate. If she wanted hard soap she "salted it out" with brine. The sodium stearate being less soluble was precipitated to the top and cooled into a solid cake that could be cut into bars by pack thread. But the frugal housewife threw away in the waste water what we now consider the most valuable ingredients, the potash and the glycerin.

But the old lye-leach is only to be found in ruins on an abandoned farm and we no longer burn wood at the

FEEDING THE SOIL

rate of a log a night. In 1916 even under the stimulus of tenfold prices the amount of potash produced as pearl ash was only 412 tons—and we need 300,000 tons in some form. It would, of course, be very desirable as a conservation measure if all the sawdust and waste wood were utilized by charring it in retorts. The gas makes a handy fuel. The tar washed from the gas contains a lot of valuable products. And potash can be leached out of the charcoal or from its ashes whenever it is burned. But this at best would not go far toward solving the problem of our national supply.

There are other potash-bearing wastes that might be utilized. The cement mills which use feldspar in combination with limestone give off a potash dust, very much to the annoyance of their neighbors. This can be collected by running the furnace clouds into large settling chambers or long flues, where the dust may be caught in bags, or washed out by water sprays or thrown down by electricity. The blast furnaces for iron also throw off potash-bearing fumes.

Our six-million-ton crop of sugar beets contains some 12,000 tons of nitrogen, 4000 tons of phosphoric acid and 18,000 tons of potash, all of which is lost except where the waste liquors from the sugar factory are used in irrigating the beet land. The beet molasses, after extracting all the sugar possible by means of lime, leaves a waste liquor from which the potash can be recovered by evaporation and charring and leaching the residue. The Germans get 5000 tons of potassium cyanide and as much ammonium sulfate annually from the waste liquor of their beet sugar factories and if it pays them to save this it ought to pay us where potash is dearer. Various other industries can put in a bit when Uncle Sam passes around the contribution basket marked "Potash for the Poor." Wool wastes and fish refuse make valuable fertilizers, although they will not go far toward solving the problem. If we saved all our potash by-products they would not supply more than fifteen per cent. of our needs.

Though no potash beds comparable to those of Stassfurt have yet been discovered in the United States, yet in Nebraska, Utah, California and other western states there are a number of alkali lakes, wet or dry, containing a considerable amount of potash mixed with soda salts. Of these deposits the largest is Searles Lake, California. Here there are some twelve square miles of salt crust some seventy feet deep and the brine as pumped out contains about four per cent. of potassium chloride. The quantity is sufficient to supply the country for over twenty years, but it is not an easy or cheap job to separate the potassium from the sodium salts which are five times more abundant. These being less soluble than the potassium salts crystallize out first when the brine is evaporated. The final crystallization is done in vacuum pans as in getting sugar from the cane juice. In this way the American Trona Corporation is producing some 4500 tons of potash salts a month besides a thousand tons of borax. The borax which is contained in the brine to the extent of 1½ per cent. is removed from the fertilizer for a double reason. It is salable by itself and it is detrimental to plant life.

Another mineral source of potash is alunite, which is a sort of natural alum, or double sulfate of potassium and aluminum, with about ten per cent. of potash. It

FEEDING THE SOIL

contains a lot of extra alumina, but after roasting in a kiln the potassium sulfate can be leached out. The alunite beds near Marysville, Utah, were worked for all they were worth during the war, but the process does not give potash cheap enough for our needs in ordinary times.

The tourist going through Wyoming on the Union Pacific will have to the north of him what is marked on the map as the "Leucite Hills." If he looks up the word in the Unabridged that he carries in his satchel he will find that leucite is a kind of lava and that it contains potash. But he will also observe that the potash is combined with alumina and silica, which are hard to get out and useless when you get them out. One of the lavas of the Leucite Hills, that named from its native state "Wyomingite," gives fifty-seven per cent. of its potash in a soluble form on roasting with alunite—but this costs too much. The same may be said of all the potash feldspars and mica. They are abundant enough, but until we find a way of utilizing the by-products, say the silica in cement and the aluminum as a metal, they cannot solve our problem.

Since it is so hard to get potash from the land it has been suggested that we harvest the sea. The experts of the United States Department of Agriculture have placed high hopes in the kelp or giant seaweed which floats in great masses in the Pacific Ocean not far off from the California coast. This is harvested with ocean reapers run by gasoline engines and brought in barges to the shore, where it may be dried and used locally as a fertilizer or burned and the potassium chloride leached out of the charcoal ashes. But it is hard to handle the bulky, slimy seaweed cheaply enough to get out of it the small amount of potash it contains. So efforts are now being made to get more out of the kelp than the potash. Instead of burning the seaweed it is fermented in vats producing acetic acid (vinegar). From the resulting liquid can be obtained lime acetate, potassium chloride, potassium iodide, acetone, ethyl acetate (used as a solvent for guncotton) and algin, a gelatin-like gum.

PRODUCTION OF POTASH IN THE UNITED STATES

Source	1916		1917	
	Tons K_2O	Per cent. of total production	Tons K_2O	Per cent. of total production
Mineral sources:				
Natural brines......	3,994	41.1	20,652	63.4
Alunite	1,850	19.0	2,402	7.3
Dust from cement mills			1,621	5.0
Dust from blast furnaces			185	0.6
Organic Sources:				
Kelp	1,556	16.0	3,752	10.9
Molasses residue from distillers	1,845	19.0	2,846	8.8
Wood ashes	412	4.2	621	1.9
Waste liquors from beet-sugar refineries			369	1.1
Miscellaneous industrial wastes	63	.7	305	1.0
Total	9,720	100.0	32,573	100.0

—From U. S. Bureau of Mines Report, 1918.

This table shows how inadequate was the reaction of the United States to the war demand for potassium salts. The minimum yearly requirements of the United States are estimated to be 250,000 tons of potash.

This completes our survey of the visible sources of potash in America. In 1917 under the pressure of the embargo and unprecedented prices the output of potash (K_2O) in various forms was raised to 32,573 tons, but this is only about a tenth as much as we needed. In 1918 potash production was further raised to 52,135 tons, chiefly through the increase of the output from natural brines to 39,255 tons, nearly twice what it was the year before. The rust in cotton and the resulting decrease in yield during the war are laid to lack of potash. Truck crops grown in soils deficient in potash do not stand transportation well. The Bureau of Animal Industry has shown in experiments in Aroostook County, Maine, that the addition of moderate amounts of potash doubled the yield of potatoes.

Professor Ostwald, the great Leipzig chemist, boasted in the war:

America went into the war like a man with a rope round his neck which is in his enemy's hands and is pretty tightly drawn. With its tremendous deposits Germany has a world monopoly in potash, a point of immense value which cannot be reckoned too highly when once this war is going to be settled. It is in Germany's power to dictate which of the nations shall have plenty of food and which shall starve.

If, indeed, some mineralogist or metallurgist will cut that rope by showing us a supply of cheap potash we will erect him a monument as big as Washington's. But Ostwald is wrong in supposing that America is as dependent as Germany upon potash. The bulk of our food crops are at present raised without the use of any fertilizers whatever.

As the cession of Lorraine in 1871 gave Germany the phosphates she needed for fertilizers so the retrocession of Alsace in 1919 gives France the potash she needed for fertilizers. Ten years before the war a bed of potash was discovered in the Forest of Monnebruck, near Hartmannsweilerkopf, the peak for which French and Germans contested so fiercely and so long. The layer of potassium salts is 16½ feet thick and the total deposit is estimated to be 275,000,000 tons of potash. At any rate it is a formidable rival of Stassfurt and its acquisition by France breaks the German monopoly.

When we turn to the consideration of the third plant food we feel better. While the United States has no such monopoly of phosphates as Germany had of potash and Chile had of nitrates we have an abundance and to spare. Whereas we formerly *imported* about $17,000,000 worth of potash from Germany and $20,000,000 worth of nitrates from Chile a year we *exported* $7,000,000 worth of phosphates.

Whoever it was who first noticed that the grass grew thicker around a buried bone he lived so long ago that we cannot do honor to his powers of observation, but ever since then—whenever it was—old bones have been used as a fertilizer. But we long ago used up all the buffalo bones we could find on the prairies and our packing houses could not give us enough bone-meal to go around, so we have had to draw upon the old boneyards of prehistoric animals. Deposits of lime phosphate of such origin were found in South Carolina in 1870 and in Florida in 1888. Since then the industry has developed with amazing rapidity until in 1913 the United States produced over three million tons of phosphates, nearly half of which was sent abroad. The chief source at present is the Florida pebbles, which are dredged up from the bottoms of lakes and rivers or washed out from the banks of streams by a hydraulic jet. The gravel is washed free from the sand and clay, screened and dried, and then is ready for shipment. The rock deposits of Florida and South Carolina are more limited than the pebble beds and may be exhausted in twenty-five or thirty years, but Tennessee and Kentucky have a lot in reserve and behind them are Idaho, Wyoming and other western states with millions of acres of phosphate land, so in this respect we are independent.

But even here the war hit us hard. For the calcium phosphate as it comes from the ground is not altogether available because it is not very soluble and the plants can only use what they can get in the water that they suck up from the soil. But if the phosphate is treated with sulfuric acid it becomes more soluble and this product is sold as "superphosphate." The sulfuric acid is made mostly from iron pyrite and this we have been content to import, over 800,000 tons of it a year, largely from Spain, although we have an abundance at home. Since the shortage of shipping shut off the foreign supply we are using more of our own pyrite and also our deposits of native sulfur along the Gulf coast. But as a consequence of this sulfuric acid during the war went up from $5 to $25 a ton and acidulated phosphates rose correspondingly.

Germany is short on natural phosphates as she is long on natural potash. But she has made up for it by utilizing a by-product of her steelworks. When phosphorus occurs in iron ore, even in minute amounts, it makes the steel brittle. Much of the iron ores of

Alsace-Lorraine were formerly considered unworkable because of this impurity, but shortly after Germany took these provinces from France in 1871 a method was discovered by two British metallurgists, Thomas and Gilchrist, by which the phosphorus is removed from the iron in the process of converting it into steel. This consists in lining the crucible or converter with lime and magnesia, which takes up the phosphorus from the melted iron. This slag lining, now rich in phosphates, can be taken out and ground up for fertilizer. So the phosphorus which used to be a detriment is now an additional source of profit and this British invention has enabled Germany to make use of the territory she stole from France to outstrip England in the steel business. In 1910 Germany produced 2,000,000 tons of Thomas slag while only 160,000 tons were produced in the United Kingdom. The open hearth process now chiefly used in the United States gives an acid instead of a basic phosphate slag, not suitable as a fertilizer. The iron ore of America, with the exception of some of the southern ores, carries so small a percentage of phosphorus as to make a basic process inadvisable.

Recently the Germans have been experimenting with a combined fertilizer, Schröder's potassium phosphate, which is said to be as good as Thomas slag for phosphates and as good as Stassfurt salts for potash. The American Cyanamid Company is just putting out a similar product, "Ammo-Phos," in which the ammonia can be varied from thirteen to twenty per cent. and the phosphoric acid from twenty to forty-seven per cent. so as to give the proportions desired for any crop. We have then the possibility of getting

FEEDING THE SOIL

the three essential plant foods altogether in one compound with the elimination of most of the extraneous elements such as lime and magnesia, chlorids and sulfates.

For the last three hundred years the American people have been living on the unearned increment of the unoccupied land. But now that all our land has been staked out in homesteads and we cannot turn to new soil when we have used up the old, we must learn, as the older races have learned, how to keep up the supply of plant food. Only in this way can our population increase and prosper. As we have seen, the phosphate question need not bother us and we can see our way clear toward solving the nitrate question. We gave the Government $20,000,000 to experiment on the production of nitrates from the air and the results will serve for fields as well as firearms. But the question of an independent supply of cheap potash is still unsolved.

IV

COAL-TAR COLORS

If you put a bit of soft coal into a test tube (or, if you have n't a test tube, into a clay tobacco pipe and lute it over with clay) and heat it you will find a gas coming out of the end of the tube that will burn with a yellow smoky flame. After all the gas comes off you will find in the bottom of the test tube a chunk of dry, porous coke. These, then, are the two main products of the destructive distillation of coal. But if you are an unusually observant person, that is, if you are a born chemist with an eye to by-products, you will notice along in the middle of the tube where it is neither too hot nor too cold some dirty drops of water and some black sticky stuff. If you are just an ordinary person, you won't pay any attention to this because there is only a little of it and because what you are after is the coke and gas. You regard the nasty, smelly mess that comes in between as merely a nuisance because it clogs up and spoils your nice, clean tube.

Now that is the way the gas-makers and coke-makers—being for the most part ordinary persons and not born chemists—used to regard the water and tar that got into their pipes. They washed it out so as to have the gas clean and then ran it into the creek. But the neighbors—especially those who fished in the stream below the gas-works—made a fuss about spoiling the water, so the gas-men gave away the tar to the boys for use in celebrating the Fourth of July and election night or sold it for roofing.

But this same tar, which for a hundred years was thrown away and nearly half of which is thrown away yet in the United States, turns out to be one of the most useful things in the world. It is one of the strategic points in war and commerce. It wounds and heals. It supplies munitions and medicines. It is like the magic purse of Fortunatus from which anything wished for could be drawn. The chemist puts his hand into the black mass and draws out all the colors of the rainbow. This evil-smelling substance beats the rose in the production of perfume and surpasses the honey-comb in sweetness.

Bishop Berkeley, after having proved that all matter was in your mind, wrote a book to prove that wood tar would cure all diseases. Nobody reads it now. The name is enough to frighten them off: "Siris: A Chain of Philosophical Reflections and Inquiries Concerning the Virtues of Tar Water." He had a sort of mystical idea that tar contained the quintessence of the forest, the purified spirit of the trees, which could somehow revive the spirit of man. People said he was crazy on the subject, and doubtless he was, but the interesting thing about it is that not even his active and ingenious imagination could begin to suggest all of the strange things that can be got out of tar, whether wood or coal.

The reason why tar supplies all sorts of useful material is because it is indeed the quintessence of the forest, of the forests of untold millenniums if it is coal tar.

If you are acquainted with a village tinker, one of those all-round mechanics who still survive in this age of specialization and can mend anything from a baby-carriage to an automobile, you will know that he has on the floor of his back shop a heap of broken machinery from which he can get almost anything he wants, a copper wire, a zinc plate, a brass screw or a steel rod. Now coal tar is the scrap-heap of the vegetable kingdom. It contains a little of almost everything that makes up trees. But you must not imagine that all that comes out of coal tar is contained in it. There are only about a dozen primary products extracted from coal tar, but from these the chemist is able to build up hundreds of thousands of new substances. This is true creative chemistry, for most of these compounds are not to be found in plants and never existed before they were made in the laboratory. It used to be thought that organic compounds, the products of vegetable and animal life, could only be produced by organized beings, that they were created out of inorganic matter by the magic touch of some "vital principle." But since the chemist has learned how, he finds it easier to make organic than inorganic substances and he is confident that he can reproduce any compound that he can analyze. He cannot only imitate the manufacturing processes of the plants and animals, but he can often beat them at their own game.

When coal is heated in the open air it is burned up and nothing but the ashes is left. But heat the coal in an enclosed vessel, say a big fireclay retort, and it cannot burn up because the oxygen of the air cannot get to it. So it breaks up. All parts of it that can be volatized at a high heat pass off through the outlet pipe and nothing is left in the retort but coke, that is carbon with the ash it contains. When the escaping vapors reach a cool part of the outlet pipe the oily and tarry matter condenses out. Then the gas is passed up through a tower down which water spray is falling and thus is washed free from ammonia and everything else that is soluble in water.

This process is called "destructive distillation." What products come off depends not only upon the composition of the particular variety of coal used, but upon the heat, pressure and rapidity of distillation. The way you run it depends upon what you are most anxious to have. If you want illuminating gas you will leave in it the benzene. If you are after the greatest yield of tar products, you impoverish the gas by taking out the benzene and get a blue instead of a bright yellow flame. If all you are after is cheap coke, you do not bother about the by-products, but let them escape and burn as they please. The tourist passing across the coal region at night could see through his car window the flames of hundreds of old-fashioned bee-hive coke-ovens and if he were of economical mind he might reflect that this display of fireworks was costing the country $75,000,000 a year besides consuming the irreplaceable fuel supply of the future. But since the gas was not needed outside of the cities and since the coal tar, if it could be sold at all, brought only a cent or two a gallon, how could the coke-makers be expected to throw out their old bee-hive ovens and put in the expensive retorts and towers necessary to the recovery of the by-products? But within the last ten years the by-product ovens have come into use and now nearly half our coke is made in them.

Although the products of destructive distillation vary within wide limits, yet the following table may serve to give an approximate idea of what may be got from a ton of soft coal:

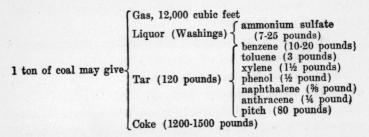

1 ton of coal may give
- Gas, 12,000 cubic feet
- Liquor (Washings)
  - ammonium sulfate (7-25 pounds)
- Tar (120 pounds)
  - benzene (10-20 pounds)
  - toluene (3 pounds)
  - xylene (1½ pounds)
  - phenol (½ pound)
  - naphthalene (⅜ pound)
  - anthracene (¼ pound)
  - pitch (80 pounds)
- Coke (1200-1500 pounds)

When the tar is redistilled we get, among other things, the ten "crudes" which are fundamental material for making dyes. Their names are: benzene, toluene, xylene, phenol, cresol, naphthalene, anthracene, methyl anthracene, phenanthrene and carbazol.

There! I had to introduce you to the whole receiving line, but now that that ceremony is over we are at liberty to do as we do at a reception, meet our old friends, get acquainted with one or two more and turn our backs on the rest. Two of them, I am sure, you've met before, phenol, which is common carbolic acid, and naphthalene, which we use for mothballs. But notice one thing in passing, that not one of them is a dye. They are all colorless liquids or white solids. Also they all have an indescribable odor—all odors that you don't know are indescribable—which gives them and their progeny, even when odorless, the name of "aromatic compounds."

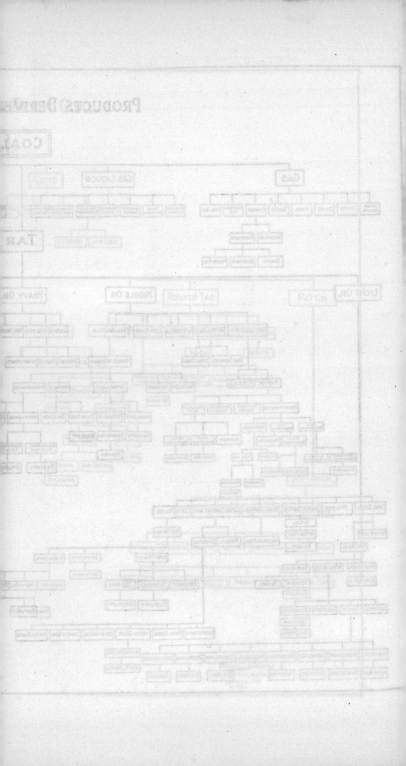

COAL-TAR COLORS

The most important of the ten because he is the father of the family is benzene, otherwise called benzol, but must not be confused with "benzine" spelled with an *i* which we used to burn and clean our clothes with. "Benzine" is a kind of gasoline, but benzene *alias* benzol has quite another constitution, although it looks and burns the same. Now the search for the constitution of benzene is one of the most exciting chapters in chemistry; also one of the most intricate chapters, but, in spite of that, I believe I can make the main point of it clear even to those who have never studied chemistry—provided they retain their childish liking for puzzles. It is really much like putting together the old six-block Chinese puzzle. The chemist can work better if he has a picture of what he is working with. Now his unit is the molecule, which is too small even to analyze with the microscope, no matter how high powered. So he makes up a sort of diagram of the molecule, and since he knows the number of atoms and that they are somehow attached to one another, he represents each atom by the first letter of its name and the points of attachment or bonds by straight lines connecting the atoms of the different elements. Now it is one of the rules of the game that all the bonds must be connected or hooked up with atoms at both ends, that there shall be no free hands reaching out into empty space. Carbon, for instance, has four bonds and hydrogen only one. They unite, therefore, in the proportion of one atom of carbon to four of hydrogen, or CH_4, which is methane or marsh gas and obviously the simplest of the hydrocarbons. But we have more complex hydrocarbons such as C_6H_{14}, known as hexane. Now if you try to draw the diagrams or structural formulas of these two compounds you will easily get

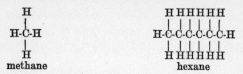

Each carbon atom, you see, has its four hands outstretched and duly grasped by one-handed hydrogen atoms or by neighboring carbon atoms in the chain. We can have such chains as long as you please, thirty or more in a chain; they are all contained in kerosene and paraffin.

So far the chemist found it easy to construct diagrams that would satisfy his sense of the fitness of things, but when he found that benzene had the composition C_6H_6 he was puzzled. If you try to draw the picture of C_6H_6 you will get something like this:

which is an absurdity because more than half of the carbon hands are waving wildly around asking to be held by something. Benzene, C_6H_6, evidently is like hexane, C_6H_{14}, in having a chain of six carbon atoms, but it has dropped its H's like an Englishman. Eight of the H's are missing.

Now one of the men who was worried over this benzene puzzle was the German chemist, Kekulé. One evening after working over the problem all day he was sitting by the fire trying to rest, but he could not throw it off his mind. The carbon and the hydrogen atoms danced like imps on the carpet and as he watched them through his half-closed eyes he suddenly saw that the chain of six carbon atoms had joined at the ends and formed a ring while the six hydrogen atoms were holding on to the outside hands, in this fashion:

$$\begin{array}{c} H \\ | \\ C \\ H-C \diagup \diagdown C-H \\ H-C \diagdown \diagup C-H \\ C \\ | \\ H \end{array}$$

Professor Kekulé saw at once that the demons of his subconscious self had furnished him with a clue to the labyrinth, and so it proved. We need not suppose that the benzene molecule if we could see it would look anything like this diagram of it, but the theory works and that is all the scientist asks of any theory. By its use thousands of new compounds have been constructed which have proved of inestimable value to man. The modern chemist is not a discoverer, he is an inventor. He sits down at his desk and draws a "Kekulé ring" or rather hexagon. Then he rubs out an H and hooks a nitro group (NO_2) on to the carbon in place of it; next he rubs out the O_2 of the nitro group and puts in H_2; then he hitches on such other elements, or carbon chains and rings as he likes. He works like an architect designing a house and when he gets a picture of the proposed compounds to suit him he goes into the laboratory to make it. First he takes down the bottle

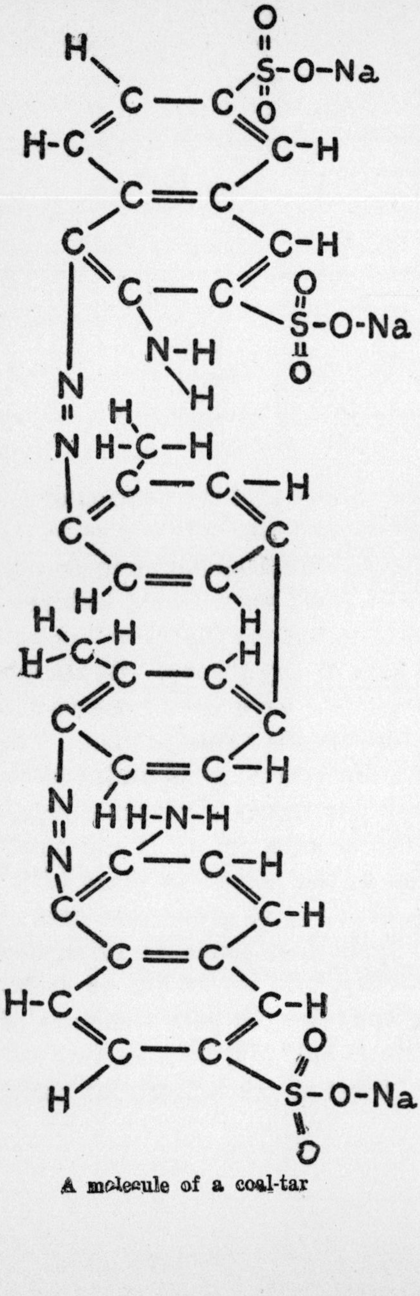

68 CREATIVE CHEMISTRY

A molecule of a coal-tar

of benzene and boils up some of this with nitric acid and sulfuric acid. This he puts in the nitro group and makes nitro-benzene, $C_6H_5\text{-}NO_2$. He treats this with hydrogen, which displaces the oxygen and gives $C_6H_5NH_2$ or aniline, which is the basis of so many of these compounds that they are all commonly called "the aniline dyes." But aniline itself is not a dye. It is a colorless or brownish oil.

It is not necessary to follow our chemist any farther now that we have seen how he works, but before we pass on we will just look at one of his products, not one of the most complicated but still complicated enough.

The name of this is sodium ditolyl-disazo-beta-naphthylamine-6-sulfonic-beta-naphthylamine-3.6-disulfonate.

These chemical names of organic compounds are discouraging to the beginner and amusing to the layman, but that is because neither of them realizes that they are not really words but formulas. They are hyphenated because they come from Germany. The name given above is no more of a mouthful than "a-square-plus-two-a-b-plus-b-square" or "Third Assistant Secretary of War to the President of the United States of America." The trade name of this dye is Brilliant Congo, but while that is handier to say it does not mean anything. Nobody but an expert in dyes would know what it was, while from the formula name any chemist familiar with such compounds could draw its picture, tell how it would behave and what it was made from, or even make it. The old alchemist was a secretive and pretentious person and used to invent queer names for the purpose of mystifying and awing the ignorant. But the chemist in dropping the al- has dropped the idea of secrecy and his names, though equally appalling to the layman, are designed to reveal and not to conceal.

From this brief explanation the reader who has not studied chemistry will, I think, be able to get some idea of how these very intricate compounds are built up step by step. A completed house is hard to understand, but when we see the mason laying one brick on top of another it does not seem so difficult, although if we tried to do it we should not find it so easy as we think. Anyhow, let me give you a hint. If you want to make a good impression on a chemist don't tell him that he seems to you a sort of magician, master of a black art,

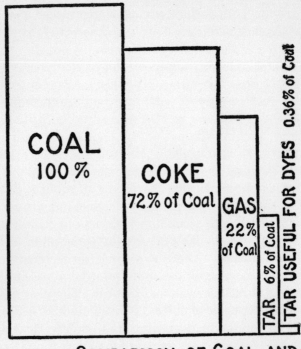

— COMPARISON OF COAL AND ITS DISTILLATION PRODUCTS

From Hesse's "The Industry of the Coal Tar Dyes," *Journal of Industrial and Engineering Chemistry*, December, 1914

and all that nonsense. The chemist has been trying for three hundred years to live down the reputation of being inspired of the devil and it makes him mad to have his past thrown up at him in this fashion. If his tactless admirers would stop saying "it is all a mystery and a miracle to me, and I cannot understand it" and pay attention to what he is telling them they would understand it and would find that it is no more of a mystery or a miracle than anything else. You can make an electrician mad in the same way by interrupting his explanation of a dynamo by asking: "But you cannot tell me what electricity really is." The electrician does not care a rap what electricity "really is"—if there really is any meaning to that phrase. All he wants to know is what he can do with it.

The tar obtained from the gas plant or the coke plant has now to be redistilled, giving off the ten "crudes" already mentioned and leaving in the still sixty-five per cent. of pitch, which may be used for roofing, paving and the like. The ten primary products or crudes are then converted into secondary products or "intermediates" by processes like that for the conversion of benzene into aniline. There are some three hundred of these intermediates in use and from them are built up more than three times as many dyes. The year before the war the American custom house listed 5674 distinct brands of synthetic dyes imported, chiefly from Germany, but some of these were trade names for the same product made by different firms or represented by different degrees of purity or form of preparation. Although the number of possible products is unlimited and over five thousand dyes are known, yet only about nine hundred are in use. We can summarize the situation so:

Coal-tar→ 10 crudes→ 300 intermediates→ 900 dyes→ 5000 brands.

Or, to borrow the neat simile used by Dr. Bernhard C.

Hesse, it is like cloth-making where "ten fibers make 300 yarns which are woven into 900 patterns."

The advantage of the artificial dyestuffs over those found in nature lies in their variety and adaptability. Practically any desired tint or shade can be made for any particular fabric. If my lady wants a new kind of green for her stockings or her hair she can have it. Candies and jellies and drinks can be made more attractive and therefore more appetizing by varied colors. Easter eggs and Easter bonnets take on new and brighter hues.

More and more the chemist is becoming the architect of his own fortunes. He does not make discoveries by picking up a beaker and pouring into it a little from each bottle on the shelf to see what happens. He generally knows what he is after, and he generally gets it, although he is still often baffled and occasionally happens on something quite unexpected and perhaps more valuable than what he was looking for. Columbus was looking for India when he ran into an obstacle that proved to be America. William Henry Perkin was looking for quinine when he blundered into that rich and undiscovered country, the aniline dyes. William Henry was a queer boy. He had rather listen to a chemistry lecture than eat. When he was attending the City of London School at the age of thirteen there was an extra course of lectures on chemistry given at the noon recess, so he skipped his lunch to take them in. Hearing that a German chemist named Hofmann had opened a laboratory in the Royal College of London he headed for that. Hofmann obviously had no fear of forcing the young intellect prematurely. He perhaps had never heard that "the tender petals of the adolescent mind must be allowed to open slowly." He admitted young Perkin at the age of fifteen and started him on research at the end of his second year. An American student nowadays thinks he is lucky if he gets started on his research five years older than Perkin. Now if Hofmann had studied pedagogical psychology he would have been informed that nothing chills the ardor of the adolescent mind like being set at tasks too great for its powers. If he had heard this and believed it, he would not have allowed Perkin to spend two years in fruitless endeavors to isolate phenanthrene from coal tar and to prepare artificial quinine— and in that case Perkin would never have discovered the aniline dyes. But Perkin, so far from being discouraged, set up a private laboratory so he could work over-time. While working here during the Easter vacation of 1856—the date is as well worth remembering as 1066—he was oxidizing some aniline oil when he got what chemists most detest, a black, tarry mass instead of nice, clean crystals. When he went to wash this out with alcohol he was surprised to find that it gave a beautiful purple solution. This was "mauve," the first of the aniline dyes.

The funny thing about it was that when Perkin tried to repeat the experiment with purer aniline he could not get his color. It was because he was working with impure chemicals, with aniline containing a little toluidine, that he discovered mauve. It was, as I said, a lucky accident. But it was not accidental that the accident happened to the young fellow who spent his noonings and vacations at the study of chemistry. A man may not find what he is looking for, but he never finds anything unless he is looking for something.

Mauve was a product of creative chemistry, for it was a substance that had never existed before. Perkin's next great triumph, ten years later, was in rivaling Nature in the manufacture of one of her own choice products. This is alizarin, the coloring matter contained in the madder root. It was an ancient and oriental dyestuff, known as "Turkey red" or by its Arabic name of "alizari." When madder was introduced into France it became a profitable crop and at one time half a million tons a year were raised. A couple of French chemists, Robiquet and Colin, extracted from madder its active principle, alizarin, in 1828, but it was not until forty years later that it was discovered that alizarin had for its base one of the coal-tar products, anthracene. Then came a neck-and-neck race between Perkin and his German rivals to see which could discover a cheap process for making alizarin from anthracene. The German chemists beat him to the patent office by one day! Graebe and Liebermann filed their application for a patent on the sulfuric acid process as No. 1936 on June 25, 1869. Perkin filed his for the same process as No. 1948 on June 26. It had required twenty years to determine the constitution of alizarin, but within six months from its first synthesis the commercial process was developed and within a few years the sale of artificial alizarin reached $8,000,000 annually. The madder fields of France were put to other uses and even the French soldiers became dependent on made-in-Germany dyes for their red trousers. The British soldiers were placed in a similar situation as regards their red coats when after 1878 the azo scarlets put the cochineal bug out of business.

The modern chemist has robbed royalty of its most distinctive insignia, Tyrian purple. In ancient times to be "porphyrogene," that is "born to the purple," was like admission to the Almanach de Gotha at the present time, for only princes or their wealthy rivals could afford to pay $600 a pound for crimsoned linen. The precious dye is secreted by a snail-like shellfish of the eastern coast of the Mediterranean. From a tiny sac behind the head a drop of thick whitish liquid, smelling like garlic, can be extracted. If this is spread upon cloth of any kind and exposed to air and sunlight it turns first green, next blue and then purple. If the cloth is washed with soap—that is, set by alkali—it becomes a fast crimson, such as Catholic cardinals still wear as princes of the church. The Phœnician merchants made fortunes out of their monopoly, but after the fall of Tyre it became one of "the lost arts"—and accordingly considered by those whose faces are set toward the past as much more wonderful than any of the new arts. But in 1909 Friedlander put an end to the superstition by analyzing Tyrian purple and finding that it was already known. It was the same as a dye that had been prepared five years before by Sachs but had not come into commercial use because of its inferiority to others in the market. It required 12,000 of the mollusks to supply the little material needed for analysis, but once the chemist had identified it he did not need to bother the Murex further, for he could make it by the ton if he had wanted to. The coloring principle turned out to be a di-brom indigo, that is the same as the substance extracted from the Indian plant, but with the additon of two atoms of bromine. Why a particular kind of a shellfish should have got the habit of extracting this rare element from sea water and stowing it away in this peculiar form is "one of those things no fellow can find out." But according to the chemist the Murex mollusk made a mistake in hitching the bromine to the wrong carbon atoms. He finds as he would word it that the 6:6′-dibrom indigo secreted by the shellfish is not so good as the 5:5′-di-brom indigo now manufactured at a cheap rate and in unlimited quantity. But we must not expect too much of a mollusk's mind. In their cheapness lies the offense of the aniline dyes in the minds of some people. Our modern aristocrats would delight to be entitled "porphyrogeniti" and to wear exclusive gowns of "purple and scarlet from the isles of Elishah" as was done in Ezekiel's time, but when any shopgirl or sailor can wear the royal color it spoils its beauty in their eyes. Applied science accomplishes a real democracy such as legislation has ever failed to establish.

Any kind of dye found in nature can be made in the laboratory whenever its composition is understood and usually it can be made cheaper and purer than it can be extracted from the plant. But to work out a profitable process for making it synthetically is sometimes a task requiring high skill, persistent labor and heavy expenditure. One of the latest and most striking of these achievements of synthetic chemistry is the manufacture of indigo.

Indigo is one of the oldest and fastest of the dyestuffs. To see that it is both ancient and lasting look

COAL-TAR COLORS

at the unfaded blue cloths that enwrap an Egyptian mummy. When Cæsar conquered our British ancestors he found them tattooed with woad, the native indigo. But the chief source of indigo was, as its name implies, India. In 1897 nearly a million acres in India were growing the indigo plant and the annual value of the crop was $20,000,000. Then the fall began and by 1914 India was producing only $300,000 worth! What had happened to destroy this profitable industry? Some blight or insect? No, it was simply that the Badische Anilin-und-Soda Fabrik had worked out a practical process for making artificial indigo.

That indigo on breaking up gave off aniline was discovered as early as 1840. In fact that was how aniline got its name, for when Fritzsche distilled indigo with caustic soda he called the colorless distillate "aniline," from the Arabic name for indigo, "anil" or "al-nil," that is, "the blue-stuff." But how to reverse the process and get indigo from aniline puzzled chemists for more than forty years until finally it was solved by Adolf von Baeyer of Munich, who died in 1917 at the age of eighty-four. He worked on the problem of the constitution of indigo for fifteen years and discovered several ways of making it. It is possible to start from benzene, toluene or naphthalene. The first process was the easiest, but if you will refer to the products of the distillation of tar you will find that the amount of toluene produced is less than the naphthalene, which is hard to dispose of. That is, if a dye factory had worked out a process for making indigo from toluene it would not be practicable because there was not enough toluene produced to supply the demand for indigo. So the more complicated napthalene process was chosen in preference to the others in order to utilize this by-product.

The Badische Anilin-und-Soda Fabrik spent $5,000,000 and seventeen years in chemical research before they could make indigo, but they gained a monopoly (or, to be exact, ninety-six per cent.) of the world's production. A hundred years ago indigo cost as much as $4 a pound. In 1914 we were paying fifteen cents a pound for it. Even the pauper labor of India could not compete with the German chemists at that price. At the beginning of the present century Germany was paying more than $3,000,000 a year for indigo. Fourteen years later Germany was *selling* indigo to the amount of $12,600,000. Besides its cheapness, artificial indigo is preferable because it is of uniform quality and greater purity. Vegetable indigo contains from forty to eighty per cent. of impurities, among them various other tinctorial substances. Artificial indigo is made pure and of any desired strength, so the dyers can depend on it.

The value of the aniline colors lies in their infinite variety. Some are fast, some will fade, some will stand wear and weather as long as the fabric, some will wash out or spot. Dyes can be made that will attach themselves to wool, to silk or to cotton, and give it any shade of any color. The period of discovery by accident has long gone by. The chemist nowadays decides first just what kind of a dye he wants, and then goes to work systematically to make it. He begins by drawing a diagram of the molecule, double-linking nitrogen or carbon and oxygen atoms to give the required intensity, putting in acid or basic radicals to fasten it to the fiber, shifting the color back and forth along the spectrum at will by introducing methyl groups, until he gets it just to his liking.

Art can go ahead of nature in the dyestuff business. Before man found that he could make all the dyes he wanted from the tar he had been burning up at home he searched the wide world over to find colors by which he could make himself—or his wife—garments as beautiful as those that arrayed the flower, the bird and the butterfly. He sent divers down into the Mediterranean to rob the murex of his purple. He sent ships to the new world to get Brazil wood and to the oldest world for indigo. He robbed the lady cochineal of her scarlet coat. Why these peculiar substances were formed only by these particular plants, mussels and insects it is hard to understand. I don't know that Mrs. Cacti Coccus derived any benefit from her scarlet uniform when khaki would be safer, and I can't imagine that to a shellfish it was of advantage to turn red as it rots or to an indigo plant that its leaves in decomposing should turn blue. But anyhow, it was man that took advantage of them until he learned how to make his own dyestuffs.

Our independent ancestors got along so far as possible with what grew in the neighborhood. Sweetapple bark gave a fine saffron yellow. Ribbons were given the hue of the rose by poke berry juice. The Confederates in their butternut-colored uniform were almost as invisible as if in khaki or *feldgrau*. Madder was cultivated in the kitchen garden. Only logwood from Jamaica and indigo from India had to be imported.

That we are not so independent today is our own fault, for we waste enough coal tar to supply ourselves and other countries with all the new dyes needed. It is essentially a question of economy and organization. We have forgotten how to economize, but we have learned how to organize.

The British Government gave the discoverer of mauve a title, but it did not give him any support in his endeavors to develop the industry, although England led the world in textiles and needed more dyes than any other country. So in 1874 Sir William Perkin relinquished the attempt to manufacture the dyes he had discovered because, as he said, Oxford and Cambridge refused to educate chemists or to carry on research. Their students, trained in the classics for the profession of being a gentleman, showed a decided repugnance to the laboratory on account of its bad smells. So when Hofmann went home he virtually took the infant industry along with him to Germany, where Ph.D.'s were cheap and plentiful and not afraid of bad smells. There the business throve amazingly, and by 1914 the Germans were manufacturing more than three-fourths of all the coal-tar products of the world and supplying material for most of the rest. The British cursed the universities for thus imperiling the nation through their narrowness and neglect; but this accusation, though natural, was not altogether fair, for at least half the blame should go to the British dyer, who did not care where his colors came from, so long as they were cheap. When finally the universities did turn over a new leaf and began to educate chemists, the manufacturers would not employ them. Before the war six English factories producing dyestuffs employed only 35 chemists altogether, while one German color works, the Höchster Farbwerke, employed 307 expert chemists and 74 technologists.

This firm united with the six other leading dye companies of Germany on January 1, 1916, to form a trust to last for fifty years. During this time they will maintain uniform prices and uniform wage scales and hours of labor, and exchange patents and secrets. They will divide the foreign business *pro rata* and share the profits. The German chemical works made big profits during the war, mostly from munitions and medicines, and will be, through this new combination, in a stronger position than ever to push the export trade.

As a consequence of letting the dye business get away from her, England found herself in a fix when war broke out. She did not have dyes for her uniforms and flags, and she did not have drugs for her wounded. She could not take advantage of the blockade to capture the German trade in Asia and South America, because she could not color her textiles. A blue cotton dyestuff that sold before the war at sixty cents a pound, brought $34 a pound. A bright pink rhodamine formerly quoted at a dollar a pound jumped to $48. When one keg of dye ordinarily worth $15 was put up at forced auction sale in 1915 it was knocked down at $1500. The Highlanders could not get the colors for their kilts until some German dyes were smuggled into England. The textile industries of Great Britain, that brought in a billion dollars a year and employed one and a half million workers, were crippled for lack of dyes. The demand for high explosives from the front could not be met because these also are largely coal-tar products. Picric acid is both a dye and an explosive. It is made from carbolic acid and the famous trinitrotoluene is made from toluene, both of which you will find in the list of the ten fundamental "crudes."

Both Great Britain and the United States realized the danger of allowing Germany to recover her former monopoly, and both have shown a readiness to cast overboard their traditional policies to meet this emergency. The British Government has discovered that a country without a tariff is a land without walls. The American Government has discovered that an industry is not benefited by being cut up into small pieces. Both governments are now doing all they can to build up big concerns and to provide them with protection. The British Government assisted in the formation of a national company for the manufacture of synthetic dyes by taking one-sixth of the stock and providing $500,000 for a research laboratory. But this effort is now reported to be "a great failure" because the Government put it in charge of the politicians instead of the chemists.

The United States, like England, had become dependent upon Germany for its dyestuffs. We imported nine-tenths of what we used and most of those that were produced here were made from imported intermediates. When the war broke out there were only seven firms and 528 persons employed in the manufacture of dyes in the United States. One of these, the Schoelkopf Aniline and Chemical Works, of Buffalo, deserves mention, for it had stuck it out ever since 1879, and in 1914 was making 106 dyes. In June, 1917, this firm, with the encouragement of the Government Bureau of Foreign and Domestic Commerce, joined with some of the other American producers to form a trade combination, the National Aniline and Chemical Company. The Du Pont Company also entered the field on an extensive scale and soon there were 118 concerns engaged in it with great profit. During the war $200,000,000 was invested in the domestic dyestuff industry. To protect this industry Congress put on a specific duty of five cents a pound and an ad valorem duty of 30 per cent. on imported dyestuffs; but if, after five years, American manufacturers are not producing 60 per cent. in value of the domestic consumption, the protection is to be removed. For some reason, not clearly understood and therefore hotly discussed, Congress at the last moment struck off the specific duty from two of the most important of the dyestuffs, indigo and alizarin, as well as from all medicinals and flavors.

The manufacture of dyes is not a big business, but it is a strategic business. Heligoland is not a big island, but England would have been glad to buy it back during the war at a high price per square yard. American industries employing over two million men and women and producing over three billion dollars' worth of products a year are dependent upon dyes. Chief of these is of course textiles, using more than half the dyes; next come leather, paper, paint and ink. We have been importing more than $12,000,000 worth of coal-tar products a year, but the cottonseed oil we exported in 1912 would alone suffice to pay that bill twice over. But although the manufacture of dyes cannot be called a big business, in comparison with some others, it is a paying business when well managed. The German concerns paid on an average 22 per cent. dividends on their capital and sometimes as high as 50 per cent. Most of the standard dyes have been so long in use that the patents are off and the processes are well enough known. We have the coal tar and we have the chemists, so there seems no good reason why we should not make our own dyes, at least enough of them so we will not be caught napping as we were in 1914. It was decidedly humiliating for our Government to have to beg Germany to sell us enough colors to print our stamps and greenbacks and then have to beg Great Britain for permission to bring them over by Dutch ships.

The raw material for the production of coal-tar products we have in abundance if we will only take the trouble to save it. In 1914 the crude light oil collected from the coke-ovens would have produced only about 4,500,000 gallons of benzol and 1,500,000 gallons of toluol, but in 1917 this output was raised to 40,200,000 gallons of benzol and 10,200,000 of toluol. The toluol was used mostly in the manufacture of trinitrotoluol for use in Europe. When the war broke out in 1914 it shut off our supply of phenol (carbolic acid) for which we were dependent upon foreign sources. This threatened not only to afflict us with headaches by depriving us of aspirin but also to remove the consolation of music, for phenol is used in making phonograph records. Mr. Edison with his accustomed energy put up a factory within a few weeks for the manufacture of synthetic phenol. When we entered the war the need for phenol became yet more imperative, for it was needed to make picric acid for filling bombs. This demand was met, and in 1917 there were fifteen new plants turning out 64,146,499 pounds of phenol valued at $23,719,805.

Some of the coal-tar products, as we see, serve many purposes. For instance, picric acid appears in three places in this book. It is a high explosive. It is a powerful and permanent yellow dye as any one who has touched it knows. Thirdly it is used as an antiseptic to cover burned skin. Other coal-tar dyes are used for the same purpose, "malachite green," "brilliant green," "crystal violet," "ethyl violet" and "Victoria blue," so a patient in a military hospital is decorated like an Easter egg. During the last five years surgeons have unfortunately had unprecedented opportunities for the study of wounds and fortunately they have been unprecedentedly successful in finding improved methods of treating them. In former wars a serious wound meant usually death or amputation. Now nearly ninety per cent. of the wounded are able to continue in the service. The reason for this improvement is that medicines are now being made to order instead of being gathered "from China to Peru." The old herb doctor picked up any strange plant that he could find and tried it on any sick man that would let him. This empirical method, though hard on the patients, resulted in the course of five thousand years in the discovery of a number of useful remedies. But the modern medicine man when he knows the cause of the disease is usually able to devise ways of counteracting it directly. For instance, he knows, thanks to Pasteur and Metchnikoff, that the cause of wound infection is the bacterial enemies of man which swarm by the million into any breach in his protective armor, the skin. Now when a breach is made in a line of intrenchments the defenders rush troops to the threatened spot for two purposes, constructive and destructive, engineers and warriors, the former to build up the rampart with sandbags, the latter to kill the enemy. So when the human body is invaded the blood brings to the breach two kinds of defenders. One is the serum which neutralizes the bacterial poison and by coagulating forms a new skin or scab over the exposed flesh. The other is the phagocytes or white corpuscles, the free lances of our corporeal militia, which attack and kill the invading bacteria. The aim of the physician then is to aid these defenders as much as possible without interfering with them. Therefore the antiseptic he is seeking is one that will assist the serum in protecting and repairing the broken tissues and will kill the hostile bacteria without killing the friendly phagocytes. Carbolic acid, the most familiar of the coal-tar antiseptics, will destroy the bacteria when it is diluted with 250 parts of water, but unfortunately it puts a stop to the fighting activities of the phagocytes when it is only half that strength, or one to 500, so it cannot destroy the infection without hindering the healing.

In this search for substances that would attack a specific disease germ one of the leading investigators was Prof. Paul Ehrlich, a German physician of the Hebrew race. He found that the aniline dyes were useful for staining slides under the microscope, for they would pick out particular cells and leave others uncolored and from this starting point he worked out

COAL-TAR COLORS 87

organic and metallic compounds which would destroy the bacteria and parasites that cause some of the most dreadful of diseases. A year after the war broke out Professor Ehrlich died while working in his laboratory on how to heal with coal-tar compounds the wounds inflicted by explosives from the same source.

One of the most valuable of the aniline antiseptics employed by Ehrlich is flavine or, if the reader prefers to call it by its full name, diaminomethylacridinium chloride. Flavine, as its name implies, is a yellow dye and will kill the germs causing ordinary abscesses when in solution as dilute as one part of the dye to 200,000 parts of water, but it does not interfere with the bactericidal action of the white blood corpuscles unless the solution is 400 times as strong as this, that is one part in 500. Unlike carbolic acid and other antiseptics it is said to stimulate the serum instead of impairing its activity. Another antiseptic of the coal-tar family which has recently been brought into use by Dr. Dakin of the Rockefeller Institute is that called by European physicians chloramine-T and by American physicians chlorazene and by chemists para-toluene-sodium-sulfo-chloramide.

This may serve to illustrate how a chemist is able to make such remedies as the doctor needs, instead of depending upon the accidental by-products of plants. On an earlier page I explained how by starting with the simplest of ring-compounds, the benzene of coal tar, we could get aniline. Suppose we go a step further and boil the aniline oil with acetic acid, which is the acid of vinegar minus its water. This easy process gives us acetanilid, which when introduced into the market some years ago under the name of "antifebrin" made a fortune for its makers.

The making of medicines from coal tar began in 1874 when Kolbe made salicylic acid from carbolic acid. Salicylic acid is a rheumatism remedy and had previously been extracted from willow bark. If now we treat salicylic acid with concentrated acetic acid we get "aspirin." From aniline again are made "phenacetin," "antipyrin" and a lot of other drugs that have become altogether too popular as headache remedies—say rather "headache relievers."

Another class of synthetics equally useful and likewise abused, are the soporifics, such as "sulphonal," "veronal" and "medinal." When it is not desired to put the patient to sleep but merely to render insensible a particular place, as when a tooth is to be pulled, cocain may be used. This, like alcohol and morphine, has proved a curse as well as a blessing and its sale has had to be restricted because of the many victims to the habit of using this drug. Cocain is obtained from the leaves of the South American coca tree, but can be made artificially from coal-tar products. The laboratory is superior to the forest because other forms of local anesthetics, such as eucain and novocain, can be made that are better than the natural alkaloid because more effective and less poisonous.

I must not forget to mention another lot of coal-tar derivatives in which some of my readers will take a personal interest. That is the photographic developers. I am old enough to remember when we used to develop our plates in ferrous sulfate solution and you never saw nicer negatives than we got with it. But when pyrogallic acid came in we switched over to that even though it did stain our fingers and sometimes our plates. Later came a swarm of new organic reducing agents under various fancy names, such as metol, hydro (short for hydro-quinone) and eikongen ("the imagemaker"). Every fellow fixed up his own formula and called his fellow-members of the camera club fools for not adopting it though he secretly hoped they would not.

Under the double stimulus of patriotism and high prices the American drug and dyestuff industry developed rapidly. In 1917 about as many pounds of dyes were manufactured in America as were imported in 1913 and our *exports* of American-made dyes exceeded in value our *imports* before the war. In 1914 the output of American dyes was valued at $2,500,000. In 1917 it amounted to over $57,000,000. This does not mean that the problem was solved, for the home products were not equal in variety and sometimes not in quality to those made in Germany. Many valuable dyes were lacking and the cost was of course much higher. Whether the American industry can compete with the foreign in an open market and on equal terms is impossible to say because such conditions did not prevail before the war and they are not going to prevail in the future. Formerly the large German cartels through their agents and branches in this country kept the business in their own hands and now the American manufacturers are determined to maintain the independence they have acquired. They will not depend hereafter upon the tariff to cut off competition but have adopted more effective measures. The 4500 German chemical patents that had been seized by the Alien Property Custodian were sold by him for $250,000 to the Chemical Foundation, an association of American manufacturers organized "for the Americanization of such institutions as may be affected thereby, for the exclusion or elimination of alien interests hostile or detrimental to said industries and for the advancement of chemical and allied science and industry in the United States." The Foundation has a large fighting fund so that it "may be able to commence immediately and prosecute with the utmost vigor infringement proceedings whenever the first German attempt shall hereafter be made to import into this country."

So much mystery has been made of the achievements of German chemists—as though the Teutonic brain had a special lobe for that faculty, lacking in other craniums—that I want to quote what Dr. Hesse says about his first impressions of a German laboratory of industrial research:

Directly after graduating from the University of Chicago in 1896, I entered the employ of the largest coal-tar dye works in the world at its plant in Germany and indeed in one of its research laboratories. This was my first trip outside the United States and it was, of course, an event of the first magnitude for me to be in Europe, and, as a chemist, to be in Germany, in a German coal-tar dye plant, and to cap it all in its research laboratory—a real *sanctum sanctorum* for chemists. In a short time the daily routine wore the novelty off my experience and I then settled down to calm analysis and dispassionate appraisal of my surroundings and to compare what was actually before and around me with my expectations.

I found that the general laboratory equipment was no better than what I had been accustomed to; that my colleagues had no better fundamental training than I had enjoyed nor any better fact—or manipulative—equipment than I; that those in charge of the work had no better general intellectual equipment nor any more native ability than had my instructors; in short, there was nothing new about it all, nothing that we did not have back home, nothing—except the specific problems that were engaging their attention, and the special opportunities of attacking them. Those problems were of no higher order of complexity than those I had been accustomed to for years, in fact, most of them were not very complex from a purely intellectual viewpoint. There was nothing inherently uncanny, magical or wizardly about their occupation whatever. It was nothing but plain hard work and keeping everlastingly at it. Now, what was the actual thing behind that chemical laboratory that we did not have at home? It was money, willing to back such activity, convinced that in the final outcome, a profit would be made; money, willing to take university graduates expecting from them no special knowledge other than a good and thorough grounding in scientific research and provide them with opportunity to become specialists suited to the factory's needs.

It is evidently not impossible to make the United States self-sufficient in the matter of coal-tar products. We 've got the tar; we 've got the men; we 've got the money, too. Whether such a policy would pay us in the long run or whether it is necessary as a measure of military or commercial self-defense is another question that cannot here be decided. But whatever share we may have in it the coal-tar industry has increased the economy of civilization and added to the wealth of the world by showing how a waste by-product could be utilized for making new dyes and valuable medicines, a better use for tar than as fuel for political bonfires and as clothing for the nakedness of social outcasts.

SYNTHETIC PERFUMES AND FLAVORS

The primitive man got his living out of such wild plants and animals as he could find. Next he, or more likely his wife, began to cultivate the plants and tame the animals so as to insure a constant supply. This was the first step toward civilization, for when men had to settle down in a community (*civitas*) they had to ameliorate their manners and make laws protecting land and property. In this settled and orderly life the plants and animals improved as well as man and returned a hundredfold for the pains that their master had taken in their training. But still man was dependent upon the chance bounties of nature. He could select, but he could not invent. He could cultivate, but he could not create. If he wanted sugar he had to send to the West Indies. If he wanted spices he had to send to the East Indies. If he wanted indigo he had to send to India. If he wanted a febrifuge he had to send to Peru. If he wanted a fertilizer he had to send to Chile. If he wanted rubber he had to send to the Congo. If he wanted rubies he had to send to Mandalay. If he wanted otto of roses he had to send to Turkey. Man was not yet master of his environment.

This period of cultivation, the second stage of civilization, began before the dawn of history and lasted until recent times. We might almost say up to the twentieth century, for it was not until the fundamental laws of heredity were discovered that man could originate new species of plants and animals according to a predetermined plan by combining such characteristics as he desired to perpetuate. And it was not until the fundamental laws of chemistry were discovered that man could originate new compounds more suitable to his purpose than any to be found in nature. Since the progress of mankind is continuous it is impossible to draw a date line, unless a very jagged one, along the frontier of human culture, but it is evident that we are just entering upon the third era of evolution in which man will make what he needs instead of trying to find it somewhere. The new epoch has hardly dawned, yet already a man may stay at home in New York or London and make his own rubber and rubies, his own indigo and otto of roses. More than this, he can make gems and colors and perfumes that never existed since time began. The man of science has signed a declaration of independence of the lower world and we are now in the midst of the revolution.

Our eyes are dazzled by the dawn of the new era. We know what the hunter and the horticulturist have already done for man, but we cannot imagine what the chemist can do. If we look ahead through the eyes of one of the greatest of French chemists, Berthelot, this is what we shall see:

> The problem of food is a chemical problem. Whenever energy can be obtained economically we can begin to make all kinds of aliment, with carbon borrowed from carbonic acid, hydrogen taken from the water and oxygen and nitrogen

drawn from the air. . . . The day will come when each person will carry for his nourishment his little nitrogenous tablet, his pat of fatty matter, his package of starch or sugar, his vial of aromatic spices suited to his personal taste; all manufactured economically and in unlimited quantities; all independent of irregular seasons, drought and rain, of the heat that withers the plant and of the frost that blights the fruit; all free from pathogenic microbes, the origin of epidemics and the enemies of human life. On that day chemistry will have accomplished a world-wide revolution that cannot be estimated. There will no longer be hills covered with vineyards and fields filled with cattle. Man will gain in gentleness and morality because he will cease to live by the carnage and destruction of living creatures. . . . The earth will be covered with grass, flowers and woods and in it the human race will dwell in the abundance and joy of the legendary age of gold —provided that a spiritual chemistry has been discovered that changes the nature of man as profoundly as our chemistry transforms material nature.

But this is looking so far into the future that we can trust no man's eyesight, not even Berthelot's. There is apparently no impossibility about the manufacture of synthetic food, but at present there is no apparent probability of it. There is no likelihood that the laboratory will ever rival the wheat field. The cornstalk will always be able to work cheaper than the chemist in the manufacture of starch. But in rarer and choicer products of nature the chemist has proved his ability to compete and even to excel.

What have been from the dawn of history to the rise of synthetic chemistry the most costly products of nature? What could tempt a merchant to brave the perils of a caravan journey over the deserts of Asia beset with Arab robbers? What induced the Portuguese and Spanish mariners to risk their frail barks on perilous waters of the Cape of Good Hope or the Horn? The chief prizes were perfumes, spices, drugs and gems. And why these rather than what now constitutes the bulk of oversea and overland commerce? Because they were precious, portable and imperishable. If the merchant got back safe after a year or two with a little flask of otto of roses, a package of camphor and a few pearls concealed in his garments his fortune was made. If a single ship of the argosy sent out from Lisbon came back with a load of sandalwood, indigo or nutmeg it was regarded as a successful venture. You know from reading the Bible, or if not that, from your reading of Arabian Nights, that a few grains of frankincense or a few drops of perfumed oil were regarded as gifts worthy the acceptance of a king or a god. These products of the Orient were equally in demand by the toilet and the temple. The unctorium was an adjunct of the Roman bathroom. Kings had to be greased and fumigated before they were thought fit to sit upon a throne. There was a theory, not yet altogether extinct, that medicines brought from a distance were most efficacious, especially if, besides being expensive, they tasted bad like myrrh or smelled bad like asafetida. And if these failed to save the princely patient he was embalmed in aromatics or, as we now call them, antiseptics of the benzene series.

Today, as always, men are willing to pay high for the titillation of the senses of smell and taste. The African savage will trade off an ivory tusk for a piece of soap reeking with synthetic musk. The clubman

SYNTHETIC PERFUMES AND FLAVORS

will pay $10 for a bottle of wine which consists mostly of water with about ten per cent. of alcohol, worth a cent or two, but contains an unweighable amount of the "bouquet" that can only be produced on the sunny slopes of Champagne or in the valley of the Rhine. But very likely the reader is quite as extravagant, for when one buys the natural violet perfumery he is paying at the rate of more than $10,000 a pound for the odoriferous oil it contains; the rest is mere water and alcohol. But you would not want the pure undiluted oil if you could get it, for it is unendurable. A single whiff of it paralyzes your sense of smell for a time just as a loud noise deafens you.

Of the five senses, three are physical and two chemical. By touch we discern pressures and surface textures. By hearing we receive impressions of certain air waves and by sight of certain ether waves. But smell and taste lead us to the heart of the molecule and enable us to tell how the atoms are put together. These twin senses stand like sentries at the portals of the body, where they closely scrutinize everything that enters. Sounds and sights may be disagreeable, but they are never fatal. A man can live in a boiler factory or in a cubist art gallery, but he cannot live in a room containing hydrogen sulfide. Since it is more important to be warned of danger than guided to delights our senses are made more sensitive to pain than pleasure. We can detect by the smell one two-millionth of a milligram of oil of roses or musk, but we can detect one two-billionth of a milligram of mercaptan, which is the vilest smelling compound that man has so far invented. If you do not know how much a milligram is consider a drop picked up by the point of a needle and imagine that divided into two billion parts. Also try to estimate the weight of the odorous particles that guide a dog to the fox or warn a deer of the presence of man. The unaided nostril can rival the spectroscope in the detection and analysis of unweighable amounts of matter.

What we call flavor or savor is a joint effect of taste and odor in which the latter predominates. There are only four tastes of importance, acid, alkaline, bitter and sweet. The acid, or sour taste, is the perception of hydrogen atoms charged with positive electricity. The alkaline, or soapy taste, is the perception of hydroxyl radicles charged with negative electricity. The bitter and sweet tastes and all the odors depend upon the chemical constitution of the compound, but the laws of the relation have not yet been worked out. Since these sense organs, the taste and smell buds, are sunk in the moist mucous membrane they can only be touched by substances soluble in water, and to reach the sense of smell they must also be volatile so as to be diffused in the air inhaled by the nose. The "taste" of food is mostly due to the volatile odors of it that creep up the back-stairs into the olfactory chamber.

A chemist given an unknown substance would have to make an elementary analysis and some tedious tests to determine whether it contained methyl or ethyl groups, whether it was an aldehyde or an ester, whether the carbon atoms were singly or doubly linked and whether it was an open chain or closed. But let him get a whiff of it and he can give instantly a pretty shrewd guess as to these points. His nose knows.

SYNTHETIC PERFUMES AND FLAVORS

Although the chemist does not yet know enough to tell for certain from looking at the structural formula what sort of odor the compound would have or whether it would have any, yet we can divide odoriferous substances into classes according to their constitution. What are commonly known as "fruity" odors belong mostly to what the chemist calls the fatty or aliphatic series. For instance, we may have in a ripe fruit an alcohol (say ethyl or common alcohol) and an acid (say acetic or vinegar)and a combination of these, the ester or organic salt (in this case ethyl acetate), which is more odorous than either of its components. These esters of the fatty acids give the characteristic savor to many of our favorite fruits, candies and beverages. The pear flavor, amyl acetate, is made from acetic acid and amyl alcohol—though amyl alcohol (fusel oil) has a detestable smell. Pineapple is ethyl butyrate—but the acid part of it (butyric acid) is what gives Limburger cheese its aroma. These essential oils are easily made in the laboratory, but cannot be extracted from the fruit for separate use.

If the carbon chain contains one or more double linkages we get the "flowery" perfumes. For instance, here is the symbol of geraniol, the chief ingredient of otto of roses:

$$(CH_3)_2C = CHCH_2CH_2C(CH_3)_2 = CHCH_2OH$$

The rose would smell as sweet under another name, but it may be questioned whether it would stand being called by the name of dimethyl-2-6-octadiene-2-6-ol-8. Geraniol by oxidation goes into the aldehyde, citral, which occurs in lemons, oranges, and verbena flowers.

Another compound of this group, linalool, is found in lavender, bergamot and many flowers.

Geraniol, as you would see if you drew up its structural formula in the way I described in the last chapter, contains a chain of six carbon atoms, that is, the same number as make a benzene ring. Now if we shake up geraniol and other compounds of this group (the diolefines) with diluted sulfuric acid the carbon chain hooks up to form a benzene ring, but with the other carbon atoms stretched across it; rather too complicated to depict here. These "bridged rings" of the formula C_5H_8, or some multiple of that, constitute the important group of the terpenes which occur in turpentine and such wild and woodsy things as sage, lavender, caraway, pine needles and eucalyptus. Going further in this direction we are led into the realm of the heavy oriental odors, patchouli, sandalwood, cedar, cubebs, ginger and camphor. Camphor can now be made directly from turpentine so we may be independent of Formosa and Borneo.

When we have a six carbon ring without double linkings (cyclo-aliphatic) or with one or two such, we get soft and delicate perfumes like the violet (ionone and irone). But when these pass into the benzene ring with its three double linkages the odor becomes more powerful and so characteristic that the name "aromatic compound" has been extended to the entire class of benzene derivatives, although many of them are odorless. The essential oils of jasmine, orange blossoms, musk, heliotrope, tuberose, ylang ylang, etc., consist mostly of this class and can be made from the common source of aromatic compounds, coal tar.

SYNTHETIC PERFUMES AND FLAVORS 101

The synthetic flavors and perfumes are made in the same way as the dyes by starting with some coal-tar product or other crude material and building up the molecule to the desired complexity. For instance, let us start with phenol, the ill-smelling and poisonous carbolic acid of disagreeable associations and evil fame. Treat this to soda-water and it is transformed into salicylic acid, a white odorless powder, used as a preservative and as a rheumatism remedy. Add to this methyl alcohol which is obtained by the destructive distillation of wood and is much more poisonous than ordinary ethyl alcohol. The alcohol and the acid heated together will unite with the aid of a little sulfuric acid and we get what the chemist calls methyl salicylate and other people call oil of wintergreen, the same as is found in wintergreen berries and birch bark. We have inherited a taste for this from our pioneer ancestors and we use it extensively to flavor our soft drinks, gum, tooth paste and candy, but the Europeans have not yet found out how nice it is.

But, starting with phenol again, let us heat it with caustic alkali and chloroform. This gives us two new compounds of the same composition, but differing a little in the order of the atoms. If you refer back to the diagram of the benzene ring which I gave in the last chapter, you will see that there are six hydrogen atoms attached to it. Now any or all these hydrogen atoms may be replaced by other elements or groups and what the product is depends not only on what the new elements are, but where they are put. It is like spelling words. The three letters t, r and a mean very different things according to whether they are put together as *art, tar* or *rat*. Or, to take a more apposite illustration, every hostess knows that the success of her dinner depends upon how she seats her guests around the table. So in the case of aromatic compounds, a little difference in the seating arrangement around the benzene ring changes the character. The two derivatives of phenol, which we are now considering, have two substituting groups. One is —O—H (called the hydroxyl group). The other is —CHO (called the aldehyde group). If these are opposite (called the para position) we have an odorless white solid. If they are side by side (called the ortho position) we have an oil with the odor of meadowsweet. Treating the odorless solid with methyl alcohol we get audepine (or anisic aldehyde) which is the perfume of hawthorn blossoms. But treating the other of the twin products, the fragrant oil, with dry acetic acid ("Perkin's reaction") we get cumarin, which is the perfume part of the tonka or tonquin beans that our forefathers used to carry in their snuff boxes. One ounce of cumarin is equal to four pounds of tonka beans. It smells sufficiently like vanilla to be used as a substitute for it in cheap extracts. In perfumery it is known as "new mown hay."

You may remember what I said on a former page about the career of William Henry Perkin, the boy who loved chemistry better than eating, and how he discovered the coal-tar dyes. Well, it is also to his ingenious mind that we owe the starting of the coal-tar perfume business which has had almost as important a development. Perkin made cumarin in 1868, but this, like the dye industry, escaped from English hands and flew over the North Sea. Before the war

Germany was exporting $1,500,000 worth of synthetic perfumes a year. Part of these went to France, where they were mixed and put up in fancy bottles with French names and sold to Americans at fancy prices.

The real vanilla flavor, vanillin, was made by Tiemann in 1874. At first it sold for nearly $800 a pound, but now it may be had for $10. How extensively it is now used in chocolate, ice cream, soda water, cakes and the like we all know. It should be noted that cumarin and vanillin, however they may be made, are not imitations, but identical with the chief constituent of the tonka and vanilla beans and, of course, are equally wholesome or harmless. But the nice palate can distinguish a richer flavor in the natural extracts, for they contain small quantities of other savory ingredients.

A true perfume consists of a large number of odoriferous chemical compounds mixed in such proportions as to produce a single harmonious effect upon the sense of smell. In a fine brand of perfume may be compounded a dozen or twenty different ingredients and these, if they are natural essences, are complex mixtures of a dozen or so distinct substances. Perfumery is one of the fine arts. The perfumer, like the orchestra leader, must know how to combine and coördinate his instruments to produce the desired sensation. A Wagnerian opera requires 103 musicians. A Strauss opera requires 112. Now if the concert manager wants to economize he will insist upon cutting down on the most expensive musicians and dropping out some of the others, say, the supernumerary violinists and the man who blows a single blast or tinkles a triangle once in the course of the evening. Only the trained ear will detect the difference and the manager can make more money.

Suppose our mercenary impresario were unable to get into the concert hall of his famous rival. He would then listen outside the window and analyze the sound in this fashion: "Fifty per cent. of the sound is made by the tuba, 20 per cent. by the bass drum, 15 per cent. by the 'cello and 10 per cent. by the clarinet. There are some other instruments, but they are not loud and I guess if we can leave them out nobody will know the difference." So he makes up his orchestra out of these four alone and many people do not know the difference.

The cheap perfumer goes about it in the same way. He analyzes, for instance, the otto or oil of roses which cost during the war $400 a pound—if you could get it at any price—and he finds that the chief ingredient is geraniol, costing only $5, and next is citronelol, costing $20; then comes nerol and others. So he makes up a cheap brand of perfumery out of three or four such compounds. But the genuine oil of roses, like other natural essences, contains a dozen or more constituents and to leave many of them out is like reducing an orchestra to a few loud-sounding instruments or a painting to a three-color print. A few years ago an attempt was made to make music electrically by producing separately each kind of sound vibration contained in the instruments imitated. Theoretically that seems easy, but practically the tone was not satisfactory because the tones and overtones of a full orchestra or even of a single violin are too numerous and complex to be reproduced individually. So the synthetic perfumes have not driven out the natural perfumes, but, on the contrary, have aided and stimulated the growth of flowers for essences. The otto or attar of roses, favorite of the Persian monarchs and romances, has in recent years come chiefly from Bulgaria. But wars are not made with rosewater and the Bulgars for the last five years have been engaged in other business than cultivating their own gardens. The alembic or still was invented by the Arabian alchemists for the purpose of obtaining the essential oil or attar of roses. But distillation, even with the aid of steam, is not altogether satisfactory. For instance, the distilled rose oil contains anywhere from 10 to 74 per cent. of a paraffin wax (stearopten) that is odorless and, on the other hand, phenyl-ethyl alcohol, which is an important constituent of the scent of roses, is broken up in the process of distillation. So the perfumer can improve on the natural or rather the distilled oil by leaving out part of the paraffin and adding the missing alcohol. Even the imported article taken direct from the still is not always genuine, for the wily Bulgar sometimes "increases the yield" by sprinkling his roses in the vat with synthetic geraniol just as the wily Italian pours a barrel of American cottonseed oil over his olives in the press.

Another method of extracting the scent of flowers is by *enfleurage,* which takes advantage of the tendency of fats to absorb odors. You know how butter set beside fish in the ice box will get a fishy flavor. In *enfleurage* moist air is carried up a tower passing alternately over trays of fresh flowers, say violets, and over glass plates covered with a thin layer of lard. The perfumed lard may then be used as a pomade or the perfume may be extracted by alcohol.

But many sweet flowers do not readily yield an essential oil, so in such cases we have to rely altogether upon more or less successful substitutes. For instance, the perfumes sold under the names of "heliotrope," "lily of the valley," "lilac," "cyclamen," "honeysuckle," "sweet pea," "arbutus," "mayflower" and "magnolia" are not produced from these flowers but are simply imitations made from other essences, synthetic or natural. Among the "thousand flowers" that contribute to the "Eau de Mille Fleurs" are the civet cat, the musk deer and the sperm whale. Some of the published formulas for "Jockey Club" call for civet or ambergris and those of "Lavender Water" for musk and civet. The less said about the origin of these three animal perfumes the better. Fortunately they are becoming too expensive to use and are being displaced by synthetic products more agreeable to a refined imagination. The musk deer may now be saved from extinction since we can make tri-nitro-butyl-xylene from coal tar. This synthetic musk passes muster to human nostrils, but a cat will turn up her nose at it. The synthetic musk is not only much cheaper than the natural, but a dozen times as strong, or let us say, goes a dozen times as far, for nobody wants it any stronger.

Such powerful scents as these are only pleasant when highly diluted, yet they are, as we have seen, essential ingredients of the finest perfumes. For instance, the natural oil of jasmine and other flowers contains traces of indols and skatols which have most disgusting odors. Though our olfactory organs cannot detect their presence yet we perceive their absence so they have to be put into the artificial perfume. Just so a brief but violent discord in a piece of music or a glaring color contrast in a painting may be necessary to the harmony of the whole.

It is absurd to object to "artificial" perfumes, for practically all perfumes now sold are artificial in the sense of being compounded by the art of the perfumer and whether the materials he uses are derived from the flowers of yesteryear or of Carboniferous Era is nobody's business but his. And he does not tell. The materials can be purchased in the open market. Various recipes can be found in the books. But every famous perfumer guards well the secret of his formulas and hands it as a legacy to his posterity. The ancient Roman family of Frangipani has been made immortal by one such hereditary recipe. The Farina family still claims to have the exclusive knowledge of how to make Eau de Cologne. This famous perfume was first compounded by an Italian, Giovanni Maria Farina, who came to Cologne in 1709. It soon became fashionable and was for a time the only scent allowed at some of the German courts. The various published recipes contain from six to a dozen ingredients, chiefly the oils of neroli, rosemary, bergamot, lemon and lavender dissolved in very pure alcohol and allowed to age like wine. The invention, in 1895, of artificial neroli (orange flowers) has improved the product.

French perfumery, like the German, had its origin in Italy, when Catherine de' Medici came to Paris as the bride of Henri II. She brought with her, among other artists, her perfumer, Sieur Toubarelli, who established himself in the flowery land of Grasse. Here for four hundred years the industry has remained rooted and the family formulas have been handed down from generation to generation. In the city of Grasse there were at the outbreak of the war fifty establishments making perfumes. The French perfumer does not confine himself to a single sense. He appeals as well to sight and sound and association. He adds to the attractiveness of his creation by a quaintly shaped bottle, an artistic box and an enticing name such as "Dans les Nues," "Le Cœur de Jeannette," "Nuit de Chine," "Un Air Embaumé," "Le Vertige," "Bon Vieux Temps," "L'Heure Bleue," "Nuit d'Amour," "Quelques Fleurs," "Djer-Kiss."

The requirements of a successful scent are very strict. A perfume must be lasting, but not strong. All its ingredients must continue to evaporate in the same proportion, otherwise it will change odor and deteriorate. Scents kill one another as colors do. The minutest trace of some impurity or foreign odor may spoil the whole effect. To mix the ingredients in a vessel of any metal but aluminum or even to filter through a tin funnel is likely to impair the perfume. The odoriferous compounds are very sensitive and unstable bodies, otherwise they would have no effect upon the olfactory organ. The combination that would be suitable for a toilet water would not be good for a talcum powder and might spoil in a soap. Perfumery is used even in the "scentless" powders and soaps. In fact it is now used more extensively, if less intensively, than ever before in the history of the world. During the Unwashed Ages, commonly called the Dark Ages, between the destruction of the Roman baths and the construction of the modern bathroom, the art of the perfumer, like all the fine arts, suffered an eclipse. "The odor of sanctity" was in highest esteem and what that odor was may be imagined from reading the lives of the saints. But in the course of centuries the refinements of life began to seep back into Europe from the East by means of the Arabs and Crusaders, and chemistry, then chiefly the art of cosmetics, began to revive. When science, the greatest democratizing agent on earth, got into action it elevated the poor to the ranks of kings and priests in the delights of the palate and the nose. We should not despise these delights, for the pleasure they confer is greater, in amount at least, than that of the so-called higher senses. We eat three times a day; some of us drink oftener; few of us visit the concert hall or the art gallery as often as we do the dining room. Then, too, these primitive senses have a stronger influence upon our emotional nature than those acquired later in the course of evolution. As Kipling puts it:

> Smells are surer than sounds or sights
> To make your heart-strings crack.

VI

CELLULOSE

Organic compounds, on which our life and living depend, consist chiefly of four elements: carbon, hydrogen, oxygen and nitrogen. These compounds are sometimes hard to analyze, but when once the chemist has ascertained their constitution he can usually make them out of their elements—if he wants to. He will not want to do it as a business unless it pays and it will not pay unless the manufacturing process is cheaper than the natural process. This depends primarily upon the cost of the crude materials. What, then, is the market price of these four elements? Oxygen and nitrogen are free as air, and as we have seen in the second chapter, their direct combination by the electric spark is possible. Hydrogen is free in the form of water but expensive to extricate by means of the electric current. But we need more carbon than anything else and where shall we get that? Bits of crystallized carbon can be picked up in South Africa and elsewhere, but those who can afford to buy them prefer to wear them rather than use them in making synthetic food. Graphite is rare and hard to melt. We must then have recourse to the compounds of carbon. The simplest of these, carbon dioxide, exists in the air but only four parts in ten thousand by volume. To extract the carbon and get it into combination with the other elements would be a difficult and expensive process. Here, then, we must call in cheap labor, the cheapest of all laborers, the plants. Pine trees on the highlands and cotton plants on the lowlands keep their green traps set all the day long and with the captured carbon dioxide build up cellulose. If, then, man wants free carbon he can best get it by charring wood in a kiln or digging up that which has been charred in nature's kiln during the Carboniferous Era. But there is no reason why he should want to go back to elemental carbon when he can have it already combined with hydrogen in the remains of modern or fossil vegetation. The synthetic products on which modern chemistry prides itself, such as vanillin, camphor and rubber, are not built up out of their elements, C, H. and O, although they might be as a laboratory stunt. Instead of that the raw material of the organic chemist is chiefly cellulose, or the products of its recent or remote destructive distillation, tar and oil.

It is unnecessary to tell the reader what cellulose is since he now holds a specimen of it in his hand, pretty pure cellulose except for the sizing and the specks of carbon that mar the whiteness of its surface. This utilization of cellulose is the chief cause of the difference between the modern world and the ancient, for what is called the invention of printing is essentially the inventing of paper. The Romans made type to stamp their coins and lead pipes with and if they had had paper to print upon the world might have escaped the Dark Ages. But the clay tablets of the Babylonians were cumbersome; the wax tablets of the Greeks were perishable; the papyrus of the Egyptians was fragile; parchment was expensive and penning was slow, so it was not until literature was put on a paper basis that democratic education became possible. At the present time sheepskin is only used for diplomas, treaties and other antiquated documents. And even if your diploma is written in Latin it is likely to be made of sulfated cellulose.

The textile industry has followed the same law of development that I have indicated in the other industries. Here again we find the three stages of progress, (1) utilization of natural products, (2) cultivation of natural products, (3) manufacture of artificial products. The ancients were dependent upon plants, animals and insects for their fibers. China used silk, Greece and Rome used wool, Egypt used flax and India used cotton. In the course of cultivation for three thousand years the animal and vegetable fibers were lengthened and strengthened and cheapened. But at last man has risen to the level of the worm and can spin threads to suit himself. He can now rival the wasp in the making of paper. He is no longer dependent upon the flax and the cotton plant, but grinds up trees to get his cellulose. A New York newspaper uses up nearly 2000 acres of forest a year. The United States grinds up about five million cords of wood a year in the manufacture of pulp for paper and other purposes.

In making "mechanical pulp" the blocks of wood, mostly spruce and hemlock, are simply pressed sidewise of the grain against wet grindstones. But in wood fiber the cellulose is in part combined with lignin, which is worse than useless. To break up the ligno-cellulose combine chemicals are used. The logs for

CELLULOSE

this are not ground fine, but cut up by disk chippers. The chips are digested for several hours under heat and pressure with acid or alkali. There are three processes in vogue. In the most common process the reagent is calcium sulfite, made by passing sulfur fumes (SO_2) into lime water. In another process a solution of caustic of soda is used to disintegrate the wood. The third, known as the "sulfate" process, should rather be called the sulfide process since the active agent is an alkaline solution of sodium sulfide made by roasting sodium sulfate with the carbonaceous matter extracted from the wood. This sulfate process, though the most recent of the three, is being increasingly employed in this country, for by means of it the resinous pine wood of the South can be worked up and the final product, known as kraft paper because it is strong, is used for wrapping.

But whatever the process we get nearly pure cellulose which, as you can see by examining this page under a microscope, consists of a tangled web of thin white fibers, the remains of the original cell walls. Owing to the severe treatment it has undergone wood pulp paper does not last so long as the linen rag paper used by our ancestors. The pages of the newspapers, magazines and books printed nowadays are likely to become brown and brittle in a few years, no great loss for the most part since they have served their purpose, though it is a pity that a few copies of the worst of them could not be printed on permanent paper for preservation in libraries so that future generations could congratulate themselves on their progress in civilization.

But in our absorption in the printed page we must not forget the other uses of paper. The paper clothing, so often prophesied, has not yet arrived. Even paper collars have gone out of fashion—if they ever were in. In Germany during the war paper was used for socks, shirts and shoes as well as handkerchiefs and napkins but it could not stand wear and washing. Our sanitary engineers have set us to drinking out of sharp-edged paper cups and we blot our faces instead of wiping them. Twine is spun of paper and furniture made of the twine, a rival of rattan. Cloth and matting woven of paper yarn are being used for burlap and grass in the making of bags and suitcases.

Here, however, we are not so much interested in manufactures of cellulose itself, that is, wood, paper and cotton, as we are in its chemical derivatives. Cellulose, as we can see from the symbol, $C_6H_{10}O_5$, is composed of the three elements of carbon, hydrogen and oxygen. These are present in the same proportion as in starch ($C_6H_{10}O_5$), while glucose or grape sugar ($C_6H_{12}O_6$) has one molecule of water more. But glucose is soluble in cold water and starch is soluble in hot, while cellulose is soluble in neither. Consequently cellulose cannot serve us for food, although some of the vegetarian animals, notably the goat, have a digestive apparatus that can handle it. In Finland and Germany birch wood pulp and straw were used not only as an ingredient of cattle food but also put into war bread. It is not likely, however, that the human stomach even under the pressure of famine is able to get much nutriment out of sawdust. But by digesting with dilute acid sawdust can be transformed into sugars and these by fermentation into alcohol, so it would be possible for a man after he has read his morning paper to get drunk on it.

If the cellulose, instead of being digested a long time in dilute acid, is dipped into a solution of sulfuric acid (50 to 80 per cent.) and then washed and dried it acquires a hard, tough and translucent coating that makes it water-proof and grease-proof. This is the "parchment paper" that has largely replaced sheepskin. Strong alkali has a similar effect to strong acid. In 1844 John Mercer, a Lancashire calico printer, discovered that by passing cotton cloth or yarn through a cold 30 per cent. solution of caustic soda the fiber is shortened and strengthened. For over forty years little attention was paid to this discovery, but when it was found that if the material was stretched so that it could not shrink on drying the twisted ribbons of the cotton fiber were changed into smooth-walled cylinders like silk, the process came into general use and nowadays much that passes for silk is "mercerized" cotton.

Another step was taken when Cross of London discovered that when the mercerized cotton was treated with carbon disulfide it was dissolved to a yellow liquid. This liquid contains the cellulose in solution as a cellulose xanthate and on acidifying or heating the cellulose is recovered in a hydrated form. If this yellow solution of cellulose is squirted out of tubes through extremely minute holes into acidulated water, each tiny stream becomes instantly solidified into a silky thread which may be spun and woven like that ejected from the spinneret of the silkworm. The origin of natural silk, if we think about it, rather detracts from the pleasure of wearing it, and if "he who needlessly sets foot upon a worm" is to be avoided as a friend we must hope that the advance of the artificial silk industry will be rapid enough to relieve us of the necessity of boiling thousands of baby worms in their cradles whenever we want silk stockings.

> On a plain rush hurdle a silkworm lay
> When a proud young princess came that way.
> The haughty daughter of a lordly king
> Threw a sidelong glance at the humble thing,
> Little thinking she walked in pride
> In the winding sheet where the silkworm died.

But so far we have not reached a stage where we can altogether dispense with the services of the silkworm. The viscose threads made by the process look as well as silk, but they are not so strong, especially when wet.

Besides the viscose method there are several other methods of getting cellulose into solution so that artificial fibers may be made from it. A strong solution of zinc chloride will serve and this process used to be employed for making the threads to be charred into carbon filaments for incandescent bulbs. Cellulose is also soluble in an ammoniacal solution of copper hydroxide. The liquid thus formed is squirted through a fine nozzle into a precipitating solution of caustic soda and glucose, which brings back the cellulose to its original form.

In the chapter on explosives I explained how cellulose treated with nitric acid in the presence of sulfuric acid was nitrated. The cellulose molecule having three hydroxyl (—OH) groups, can take up one, two or three

CELLULOSE

nitrate groups ($-ONO_2$). The higher nitrates are known as guncotton and form the basis of modern dynamite and smokeless powder. The lower nitrates, known as pyroxylin, are less explosive, although still very inflammable. All these nitrates are, like the original cellulose, insoluble in water, but unlike the original cellulose, soluble in a mixture of ether and alcohol. The solution is called collodion and is now in common use to spread a new skin over a wound. The great war might be traced back to Nobel's cut finger. Alfred Nobel was a Swedish chemist—and a pacifist. One day while working in the laboratory he cut his finger, as chemists are apt to do, and, again as chemists are apt to do, he dissolved some guncotton in ether-alcohol and swabbed it on the wound. At this point, however, his conduct diverges from the ordinary, for instead of standing idle, impatiently waving his hand in the air to dry the film as most people, including chemists, are apt to do, he put his mind on it and it occurred to him that this sticky stuff, slowly hardening to an elastic mass, might be just the thing he was hunting as an absorbent and solidifier of nitroglycerin. So instead of throwing away the extra collodion that he had made he mixed it with nitroglycerin and found that it set to a jelly. The "blasting gelatin" thus discovered proved to be so insensitive to shock that it could be safely transported or fired from a cannon. This was the first of the high explosives that have been the chief factor in modern warfare.

But on the whole, collodion has healed more wounds than it has caused besides being of infinite service to mankind otherwise. It has made modern photography possible, for the film we use in the camera and moving picture projector consists of a gelatin coating on a pyroxylin backing. If collodion is forced through fine glass tubes instead of through a slit, it comes out a thread instead of a film. If the collodion jet is run into a vat of cold water the ether and alcohol dissolve; if it is run into a chamber of warm air they evaporate. The thread of nitrated cellulose may be rendered less inflammable by taking out the nitrate groups by treatment with ammonium or calcium sulfide. This restores the original cellulose, but now it is an endless thread of any desired thickness, whereas the native fiber was in size and length adapted to the needs of the cottonseed instead of the needs of man. The old motto, "If you want a thing done the way you want it you must do it yourself," explains why the chemist has been called in to supplement the work of nature in catering to human wants.

Instead of nitric acid we may use strong acetic acid to dissolve the cotton. The resulting cellulose acetates are less inflammable than the nitrates, but they are more brittle and more expensive. Motion picture films made from them can be used in any hall without the necessity of imprisoning the operator in a fire-proof box where if anything happens he can burn up all by himself without disturbing the audience. The cellulose acetates are being used for auto goggles and gas masks as well as for windows in leather curtains and transparent coverings for index cards. A new use that has lately become important is the varnishing of aeroplane wings, as it does not readily absorb water or catch fire and makes the cloth taut and air-tight. Aeroplane wings can be made of cellulose acetate sheets as transparent as those of a dragon-fly and not easy to see against the sky.

The nitrates, sulfates and acetates are the salts or esters of the respective acids, but recently true ethers or oxides of cellulose have been prepared that may prove still better since they contain no acid radicle and are neutral and stable.

These are in brief the chief processes for making what is commonly but quite improperly called "artificial silk." They are not the same substance as silkworm silk and ought not to be—though they sometimes are—sold as such. They are none of them as strong as the silk fiber when wet, although if I should venture to say which of the various makes weakens the most on wetting I should get myself into trouble. I will only say that if you have a grudge against some fisherman give him a fly line of artificial silk, 'most any kind.

The nitrate process was discovered by Count Hilaire de Chardonnet while he was at the Polytechnic School of Paris, and he devoted his life and his fortune trying to perfect it. Samples of the artificial silk were exhibited at the Paris Exposition in 1889 and two years later he started a factory at Basançon. In 1892, Cross and Bevan, English chemists, discovered the viscose or xanthate process, and later the acetate process. But although all four of these processes were invented in France and England, Germany reaped most benefit from the new industry, which was bringing into that country $6,000,000 a year before the war. The largest producer in the world was the Vereinigte Glanzstoff-Fabriken of Elberfeld, which was paying annual dividends of 34 per cent. in 1914.

The raw materials, as may be seen, are cheap and abundant, merely cellulose, salt, sulfur, carbon, air and water. Any kind of cellulose can be used, cotton waste, rags, paper, or even wood pulp. The processes are various, the names of the products are numerous and the uses are innumerable. Even the most inattentive must have noticed the widespread employment of these new forms of cellulose. We can buy from a street barrow for fifteen cents near-silk neckties that look as well as those sold for seventy-five. As for wear—well, they all of them wear till after we get tired of wearing them. Paper "vulcanized" by being run through a 30 per cent. solution of zinc chloride and subjected to hydraulic pressure comes out hard and horny and may be used for trunks and suit cases. Viscose tubes for sausage containers are more sanitary and appetizing than the customary casings. Viscose replaces ramie or cotton in the Welsbach gas mantles. Viscose film, transparent and a thousandth of an inch thick (cellophane), serves for candy wrappers. Cellulose acetate cylinders spun out of larger orifices than silk are trying—not very successfully as yet—to compete with hog's bristles and horsehair. Stir powdered metals into the cellulose solution and you have the Bayko yarn. Bayko (from the manufacturers, Farbenfabriken vorm. Friedr. Bayer and Company) is one of those telescoped names like Socony, Nylic, Fominco, Alco, Ropeco, Ripans, Penn-Yan, Anzac, Dagor, Dora and Cadets, which will be the despair of future philologers.

Soluble cellulose may enable us in time to dispense with the weaver as well as the silk-worm. It may by one operation give us fabrics instead of threads. A machine has been invented for manufacturing net and lace, the liquid material being poured on one side of a roller and the fabric being reeled off on the other side. The process seems capable of indefinite extension and application to various sorts of woven, knit and reticulated goods. The raw material is cotton waste and the finished fabric is a good substitute for silk. As in the process of making artificial silk the cellulose is dissolved in a cupro-ammoniacal solution, but instead of being forced out through minute openings to form threads, as in that process, the paste is allowed to flow upon a revolving cylinder which is engraved with the pattern of the desired textile. A scraper removes the excess and the turning of the cylinder brings the paste in the engraved lines down into a bath which solidifies it.

Tulle or net is now what is chiefly being turned out, but the engraved design may be as elaborate and artistic as desired, and various materials can be used. Since the threads wherever they cross are united, the fabric is naturally stronger than the ordinary. It is all of a piece and not composed of parts. In short, we seem to be on the eve of a revolution in textiles that is the same as that taking place in building materials. Our concrete structures, however great, are all one stone. They are not built up out of blocks, but cast as a whole.

Lace has always been the aristocrat among textiles. It has maintained its exclusiveness hitherto by being based upon hand labor. In no other way could one get so much painful, patient toil put into such a light and portable form. A filmy thing twined about a neck or dropping from a wrist represented years of work by poor peasant girls or pallid, unpaid nuns. A visit to a lace factory, even to the public rooms where the worn-out women were not to be seen, is enough to make one resolve never to purchase any such thing made by hand again. But our good resolutions do not last long and in time we forget the strained eyes and bowed backs, or, what is worse, value our bit of lace all the more because it means that some poor woman has put her life and health into it, netting and weaving, purling and knotting, twining and twisting, throwing and drawing, thread by thread, day after day, until her eyes can no longer see and her fingers have become stiffened.

But man is not naturally cruel. He does not really enjoy being a slave driver, either of human or animal slaves, although he can be hardened to it with shocking ease if there seems no other way of getting what he wants. So he usually welcomes that Great Liberator, the Machine. He prefers to drive the tireless engine than to whip the straining horses. He had rather see the farmer riding at ease in a mowing machine than bending his back over a scythe.

The Machine is not only the Great Liberator, it is the Great Leveler also. It is the most powerful of the forces for democracy. An aristocracy can hardly be maintained except by distinction in dress, and distinction in dress can only be maintained by sumptuary laws or costliness. Sumptuary laws are unconstitutional in this country, hence the stress laid upon costliness.

But machinery tends to bring styles and fabrics within the reach of all. The shopgirl is almost as well dressed on the street as her rich customer. The man who buys ready-made clothing is only a few weeks behind the vanguard of the fashion. There is often no difference perceptible to the ordinary eye between cheap and high-priced clothing once the price tag is off. Jewels as a portable form of concentrated costliness have been in favor from the earliest ages, but now they are losing their factitious value through the advance of invention. Rubies of unprecedented size, not imitation, but genuine rubies, can now be manufactured at reasonable rates. And now we may hope that lace may soon be within the reach of all, not merely lace of the established forms, but new and more varied and intricate and beautiful designs, such as the imagination has been able to conceive, but the hand cannot execute.

Dissolving nitrocellulose in ether and alcohol we get the collodion varnish that we are all familiar with since we have used it on our cut fingers. Spread it on cloth instead of your skin and it makes a very good leather substitute. As we all know to our cost the number of animals to be skinned has not increased so rapidly in recent years as the number of feet to be shod. After having gone barefoot for a million years or so the majority of mankind have decided to wear shoes and this change in fashion comes at a time, roughly speaking, when pasture land is getting scarce. Also there are books to be bound and other new things to be done for which leather is needed. The war has intensified the stringency; so has feminine fashion. The conventions require that the shoe-tops extend nearly to skirt-bottom and this means that an inch or so must be added to the shoe-top every year. Consequent to this rise in leather we have to pay as much for one shoe as we used to pay for a pair.

Here, then, is a chance for Necessity to exercise her maternal function. And she has responded nobly. A progeny of new substances have been brought forth and, what is most encouraging to see, they are no longer trying to worm their way into favor as surreptitious surrogates under the names of "leatheret," "leatherine," "leatheroid" and "leather-this-or-that" but come out boldly under names of their own coinage and declare themselves not an imitation, not even a substitute, but "better than leather." This policy has had the curious result of compelling the cowhide men to take full pages in the magazines to call attention to the forgotten virtues of good old-fashioned sole-leather! There are now upon the market synthetic shoes that a vegetarian could wear with a clear conscience. The soles are made of some rubber composition; the uppers of cellulose fabric (canvas) coated with a cellulose solution such as I have described.

Each firm keeps its own process for such substance a dead secret, but without prying into these we can learn enough to satisfy our legitimate curiosity. The first of the artificial fabrics was the old-fashioned and still indispensable oil-cloth, that is canvas painted or printed with linseed oil carrying the desired pigments. Linseed oil belongs to the class of compounds that the chemist calls "unsaturated" and the psychologist would call "unsatisfied." They take up oxygen from the air and become solid, hence are called the "drying oils," although this does not mean that they lose water, for they have not any to lose. Later, ground cork was mixed with the linseed oil and then it went by its Latin name, "linoleum."

The next step was to cut loose altogether from the natural oils and use for the varnish a solution of some of the cellulose esters, usually the nitrate (pyroxylin or guncotton), more rarely the acetate. As a solvent the ether-alcohol mixture forming collodion was, as we have seen, the first to be employed, but now various other solvents are in use, among them castor oil, methyl alcohol, acetone, and the acetates of amyl or ethyl. Some of these will be recognized as belonging to the fruit essences that we considered in Chapter V, and doubtless most of us have perceived an odor as of overripe pears, bananas or apples mysteriously emanating from a newly lacquered radiator. With powdered bronze, imitation gold, aluminum or something of the kind a metallic finish can be put on any surface.

Canvas coated or impregnated with such soluble cellulose gives us new flexible and durable fabrics that have other advantages over leather besides being cheaper and more abundant. Without such material for curtains and cushions the automobile business would have been sorely hampered. It promises to provide us with a book binding that will not crumble to powder in the course of twenty years. Linen collars may be water-proofed and possibly Dame Fashion—being a fickle lady—may some day relent and let us wear such sanitary and economical neckwear. For shoes, purses, belts and the like the cellulose varnish or veneer is usually colored and stamped to resemble the grain of any kind of leather desired, even snake or alligator.

If instead of dissolving the cellulose nitrate and spreading it on fabric we combine it with camphor we get celluloid, a plastic solid capable of innumerable applications. But that is another story and must be reserved for the next chapter.

But before leaving the subject of cellulose proper I must refer back again to its chief source, wood. We inherited from the Indians a well-wooded continent. But the pioneer carried an ax on his shoulder and began using it immediately. For three hundred years the trees have been cut down faster than they could grow, first to clear the land, next for fuel, then for lumber and lastly for paper. Consequently we are within sight of a shortage of wood as we are of coal and oil. But the coal and oil are irrecoverable while the wood may be regrown, though it would require another three hundred years and more to grow some of the trees we have cut down. For fuel a pound of coal is about equal to two pounds of wood, and a pound of gasoline to three pounds of wood in heating value, so there would be a great loss in efficiency and economy if the world had to go back to a wood basis. But when that time shall come, as, of course, it must come some time, the wood will doubtless not be burned in its natural state but will be converted into hydrogen and carbon monoxide in a gas producer or will be distilled in closed ovens giving charcoal and gas and saving the by-products, the tar and acid liquors. As it is now the lumberman wastes two-thirds of every tree he cuts down. The rest is left in the forest as stump and tops

CELLULOSE

or thrown out at the mill as sawdust and slabs. The slabs and other scraps may be used as fuel or worked up into small wood articles like laths and clothes-pins. The sawdust is burned or left to rot. But it is possible, although it may not be profitable, to save all this waste.

In a former chapter I showed the advantages of the introduction of by-product coke-ovens. The same principle applies to wood as to coal. If a cord of wood (128 cubic feet) is subjected to a process of destructive distillation it yields about 50 bushels of charcoal, 11,500 cubic feet of gas, 25 gallons of tar, 10 gallons of crude wood alcohol and 200 pounds of crude acetate of lime. Resinous woods such as pine and fir distilled with steam give turpentine and rosin. The acetate of lime gives acetic acid and acetone. The wood (methyl) alcohol is almost as useful as grain (ethyl) alcohol in arts and industry and has the advantage of killing off those who drink it promptly instead of slowly.

The chemist is an economical soul. He is never content until he has converted every kind of waste product into some kind of profitable by-product. He now has his glittering eye fixed upon the mountains of sawdust that pile up about the lumber mills. He also has a notion that he can beat lumber for some purposes.

VII

SYNTHETIC PLASTICS

In the last chapter I told how Alfred Nobel cut his finger and, daubing it over with collodion, was led to the discovery of high explosive, dynamite. I remarked that the first part of this process—the hurting and the healing of the finger—might happen to anybody but not everybody would be led to discovery thereby. That is true enough, but we must not think that the Swedish chemist was the only observant man in the world. About this same time a young man in Albany, named John Wesley Hyatt, got a sore finger and resorted to the same remedy and was led to as great a discovery. His father was a blacksmith and his education was confined to what he could get at the seminary of Eddytown, New York, before he was sixteen. At that age he set out for the West to make his fortune. He made it, but after a long, hard struggle. His trade of typesetter gave him a living in Illinois, New York or wherever he wanted to go, but he was not content with his wages or his hours. However, he did not strike to reduce his hours or increase his wages. On the contrary, he increased his working time and used it to increase his income. He spent his nights and Sundays in making billiard balls, not at all the sort of thing you would expect of a young man of his Christian name. But working with billiard balls is more profitable than play-

SYNTHETIC PLASTICS 129

ing with them—though that is not the sort of thing you would expect a man of my surname to say. Hyatt had seen in the papers an offer of a prize of $10,000 for the discovery of a satisfactory substitute for ivory in the making of billiard balls and he set out to get that prize. I don't know whether he ever got it or not, but I have in my hand a newly published circular announcing that Mr. Hyatt has now perfected a process for making billiard balls "better than ivory." Meantime he has turned out several hundred other inventions, many of them much more useful and profitable, but I imagine that he takes less satisfaction in any of them than he does in having solved the problem that he undertook fifty years ago.

The reason for the prize was that the game on the billiard table was getting more popular and the game in the African jungle was getting scarcer, especially elephants having tusks more than 2 7/16 inches in diameter. The raising of elephants is not an industry that promises as quick returns as raising chickens or Belgian hares. To make a ball having exactly the weight, color and resiliency to which billiard players have become accustomed seemed an impossibility. Hyatt tried compressed wood, but while he did not succeed in making billiard balls he did build up a profitable business in stamped checkers and dominoes.

Setting type in the way they did it in the sixties was hard on the hands. And if the skin got worn thin or broken the dirty lead type were liable to infect the fingers. One day in 1863 Hyatt, finding his fingers were getting raw, went to the cupboard where was kept the "liquid cuticle" used by the printers. But when he got there he found it was bare, for the vial had tipped over—you know how easily they tip over—and the collodion had run out and solidified on the shelf. Possibly Hyatt was annoyed, but if so he did not waste time raging around the office to find out who tipped over that bottle. Instead he pulled off from the wood a bit of the dried film as big as his thumb nail and examined it with that " 'satiable curtiosity," as Kipling calls it, which is characteristic of the born inventor. He found it tough and elastic and it occurred to him that it might be worth $10,000. It turned out to be worth many times that.

Collodion, as I have explained in previous chapters, is a solution in ether and alcohol of guncotton (otherwise known as pyroxylin or nitrocellulose), which is made by the action of nitric acid on cotton. Hyatt tried mixing the collodion with ivory powder, also using it to cover balls of the necessary weight and solidity, but they did not work very well and besides were explosive. A Colorado saloon keeper wrote in to complain that one of the billiard players had touched a ball with a lighted cigar, which set it off and every man in the room had drawn his gun.

The trouble with the dissolved guncotton was that it could not be molded. It did not swell up and set; it merely dried up and shrunk. When the solvent evaporated it left a wrinkled, shriveled, horny film, satisfactory to the surgeon but not to the man who wanted to make balls and hairpins and knife handles out of it. In England Alexander Parkes began working on the problem in 1855 and stuck to it for ten years before he, or rather his backers, gave up. He tried mixing in

SYNTHETIC PLASTICS

various things to stiffen up the pyroxylin. Of these, camphor, which he tried in 1865, worked the best, but since he used castor oil to soften the mass articles made of "parkesine" did not hold up in all weathers.

Another Englishman, Daniel Spill, an associate of Parkes, took up the problem where he had dropped it and turned out a better product, "xylonite," though still sticking to the idea that castor oil was necessary to get the two solids, the guncotton and the camphor, together.

But Hyatt, hearing that camphor could be used and not knowing enough about what others had done to follow their false trails, simply mixed his camphor and guncotton together without any solvent and put the mixture in a hot press. The two solids dissolved one another and when the press was opened there was a clear, solid, homogeneous block of—what he named— "celluloid." The problem was solved and in the simplest imaginable way. Tissue paper, that is, cellulose, is treated with nitric acid in the presence of sulfuric acid. The nitration is not carried so far as to produce the guncotton used in explosives but only far enough to make a soluble nitrocellulose or pyroxylin. This is pulped and mixed with half the quantity of camphor, pressed into cakes and dried. If this mixture is put into steam-heated molds and subjected to hydraulic pressure it takes any desired form. The process remains essentially the same as was worked out by the Hyatt brothers in the factory they set up in Newark in 1872 and some of their original machines are still in use. But this protean plastic takes innumerable forms and almost as many names. Each factory has its own secrets and lays claim to peculiar merits. The fundamental product itself is not patented, so trade names are copyrighted to protect the product. I have already mentioned three, "parkesine," "xylonite" and "celluloid," and I may add, without exhausting the list of species belonging to this genus, "viscoloid," "lithoxyl," "fiberloid," "coraline," "eburite," "pulveroid," "ivorine," "pergamoid," "duroid," "ivortus," "crystalloid," "transparene," "litnoid," "petroid," "pasbosene," "cellonite" and "pyralin."

Celluloid can be given any color or colors by mixing in aniline dyes or metallic pigments. The color may be confined to the surface or to the interior or pervade the whole. If the nitrated tissue paper is bleached the celluloid is transparent or colorless. In that case it is necessary to add an antacid such as urea to prevent its getting yellow or opaque. To make it opaque and less inflammable oxides or chlorides of zinc, aluminum, magnesium, etc., are mixed in.

Without going into the question of their variations and relative merits we may consider the advantages of the pyroxylin plastics in general. Here we have a new substance, the product of the creative genius of man, and therefore adaptable to his needs. It is hard but light, tough but elastic, easily made and tolerably cheap. Heated to the boiling point of water it becomes soft and flexible. It can be turned, carved, ground, polished, bent, pressed, stamped, molded or blown. To make a block of any desired size simply pile up the sheets and put them in a hot press. To get sheets of any desired thickness, simply shave them off the block. To make a tube of any desired size, shape or thickness

SYNTHETIC PLASTICS

squirt out the mixture through a ring-shaped hole or roll the sheets around a hot bar. Cut the tube into sections and you have rings to be shaped and stamped into box bodies or napkin rings. Print words or pictures on a celluloid sheet, put a thin transparent sheet over it and weld them together, then you have something like the horn book of our ancestors, but better.

Nowadays such things as celluloid and pyralin can be sold under their own name, but in the early days the artificial plastics, like every new thing, had to resort to *camouflage,* a very humiliating expedient since in some cases they were better than the material they were forced to imitate. Tortoise shell, for instance, cracks, splits and twists, but a "tortoise shell" comb of celluloid looks as well and lasts better. Horn articles are limited to size of the ceratinous appendages that can be borne on the animal's head, but an imitation of horn can be made of any thickness by wrapping celluloid sheets about a cone. Ivory, which also has a laminated structure, may be imitated by rolling together alternate white opaque and colorless translucent sheets. Some of the sheets are wrinkled in order to produce the knots and irregularities of the grain of natural ivory. Man's chief difficulty in all such work is to imitate the imperfections of nature. His whites are too white, his surfaces are too smooth, his shapes are too regular, his products are too pure.

The precious red coral of the Mediterranean can be perfectly imitated by taking a cast of a coral branch and filling in the mold with celluloid of the same color and hardness. The clear luster of amber, the dead black of ebony, the cloudiness of onyx, the opalescence of alabaster, the glow of carnelian—once confined to the selfish enjoyment of the rich—are now within the reach of every one, thanks to this chameleon material. Mosaics may be multiplied indefinitely by laying together sheets and sticks of celluloid, suitably cut and colored to make up the picture, fusing the mass, and then shaving off thin layers from the end. That *chef d'œuvre* of the Venetian glass makers, the Battle of Isus, from the House of the Faun in Pompeii, can be reproduced as fast as the machine can shave them off the block. And the tesserae do not fall out like those you bought on the Rialto.

The process thus does for mosaics, ivory and coral what printing does for pictures. It is a mechanical multiplier and only by such means can we ever attain to a state of democratic luxury. The product, in cases where the imitation is accurate, is equally valuable except to those who delight in thinking that coral insects, Italian craftsmen and elephants have been laboring for years to put a trinket into their hands. The Lord may be trusted to deal with such selfish souls according to their deserts.

But it is very low praise for a synthetic product that it can pass itself off, more or less acceptably, as a natural product. If that is all we could do without it. It must be an improvement in some respects on anything to be found in nature or it does not represent a real advance. So celluloid and its congeners are not confined to the shapes of shell and coral and crystal, or to the grain of ivory and wood and horn, the colors of amber and amethyst and lapis lazuli, but can be given

SYNTHETIC PLASTICS

forms and textures and tints that were never known before 1869.

Let me see now, have I mentioned all the uses of celluloid? Oh, no, there are handles for canes, umbrellas, mirrors and brushes, knives, whistles, toys, blown animals, card cases, chains, charms, brooches, badges, bracelets, rings, book bindings, hairpins, campaign buttons, cuff and collar buttons, cuffs, collars and dickies, tags, cups, knobs, paper cutters, picture frames, chessmen, pool balls, ping pong balls, piano keys, dental plates, masks for disfigured faces, penholders, eyeglass frames, goggles, playing cards—and you can carry on the list as far as you like.

Celluloid has its disadvantages. You may mold, you may color the stuff as you will, the scent of the camphor will cling around it still. This is not usually objectionable except where the celluloid is trying to pass itself off for something else, in which case it deserves no sympathy. It is attacked and dissolved by hot acids and alkalies. It softens up when heated, which is handy in shaping it though not so desirable afterward. But the worst of its failings is its combustibility. It is not explosive, but it takes fire from a flame and burns furiously with clouds of black smoke.

But celluloid is only one of many plastic substances that have been introduced to the present generation. A new and important group of them is now being opened up, the so-called "condensation products." If you will take down any old volume of chemical research you will find occasionally words to this effect: "The reaction resulted in nothing but an insoluble resin which was not further investigated." Such a passage would be marked with a tear if chemists were given to crying over their failures. For it is the epitaph of a buried hope. It likely meant the loss of months of labor. The reason the chemist did not do anything further with the gummy stuff that stuck up his test tube was because he did not know what to do with it. It could not be dissolved, it could not be crystallized, it could not be distilled, therefore it could not be purified, analyzed and identified.

What had happened was in most cases this. The molecule of the compound that the chemist was trying to make had combined with others of its kind to form a molecule too big to be managed by such means. Financiers call the process a "merger." Chemists call it "polymerization." The resin was a molecular trust, indissoluble, uncontrollable and contaminating everything it touched.

But chemists—like governments—have learned wisdom in recent years. They have not yet discovered in all cases how to undo the process of polymerization, or, if you prefer the financial phrase, how to unscramble the eggs. But they have found that these molecular mergers are very useful things in their way. For instance there is a liquid known as isoprene (C_5H_8). This on heating or standing turns into a gum, that is nothing less than rubber, which is some multiple of C_5H_8.

For another instance there is formaldehyde, an acrid smelling gas, used as a disinfectant. This has the simplest possible formula for a carbohydrate, CH_2O. But in the leaf of a plant this molecule multiplies itself by six and turns into a sweet solid glucose ($C_6H_{12}O_6$), or with the loss of water into starch ($C_6H_{10}O_5$) or cellulose ($C_6H_{10}O_5$).

But formaldehyde is so insatiate that it not only combines with itself but seizes upon other substances, particularly those having an acquisitive nature like its own. Such a substance is carbolic acid (phenol) which, as we all know, is used as a disinfectant like formaldehyde because it, too, has the power of attacking decomposable organic matter. Now Prof. Adolf von Baeyer discovered in 1872 that when phenol and formaldehyde were brought into contact they seized upon one another and formed a combine of unusual tenacity, that is, a resin. But as I have said, chemists in those days were shy of resins. Kleeberg in 1891 tried to make something out of it and W. H. Story in 1895 went so far as to name the product "resinite," but nothing came of it until 1909 when L. H. Baekeland undertook a serious and systematic study of this reaction in New York. Baekeland was a Belgian chemist, born at Ghent in 1863 and professor at Bruges. While a student at Ghent he took up photography as a hobby and began to work on the problem of doing away with the dark-room by producing a printing paper that could be developed under ordinary light. When he came over to America in 1889 he brought his idea with him and four years later turned out "Velox," with which doubtless the reader is familiar. Velox was never patented because, as Dr. Baekeland explained in his speech of acceptance of the Perkin medal from the chemists of America, lawsuits are too expensive. Manufacturers seem to be coming generally to the opinion that a synthetic name copyrighted as a trademark affords better protection than a patent.

Later Dr. Baekeland turned his attention to the phenol condensation products, working gradually up from test tubes to ton vats according to his motto: "Make your mistakes on a small scale and your profits on a large scale." He found that when equal weights of phenol and formaldehyde were mixed and warmed in the presence of an alkaline catalytic agent the solution separated into two layers, the upper aqueous and the lower a resinous precipitate. This resin was soft, viscous and soluble in alcohol or acetone. But if it was heated under pressure it changed into another and a new kind of resin that was hard, inelastic, unplastic, infusible and insoluble. The chemical name of this product is "polymerized oxybenzyl methylene glycol anhydride," but nobody calls it that, not even chemists. It is called "Bakelite" after its inventor.

The two stages in its preparation are convenient in many ways. For instance, porous wood may be soaked in the soft resin and then by heat and pressure it is changed to the bakelite form and the wood comes out with a hard finish that may be given the brilliant polish of Japanese lacquer. Paper, cardboard, cloth, wood pulp, sawdust, asbestos and the like may be impregnated with the resin, producing tough and hard material suitable for various purposes. Brass work painted with it and then baked at 300° F. acquires a lacquered surface that is unaffected by soap. Forced in powder or sheet form into molds under a pressure of 1200 to 2000 pounds to the square inch it takes the most delicate impressions. Billiard balls of bakelite are claimed to be better than ivory because, having no grain, they do not swell unequally with heat and humidity and so lose their sphericity. Pipestems and beads of bakelite have the clear brilliancy of amber and greater strength. Fountain pens made of it are transparent so you can see how much ink you have left. A new and enlarging field for bakelite and allied products is the making of noiseless gears for automobiles and other machinery, also of air-plane propellers.

Celluloid is more plastic and elastic than bakelite. It is therefore more easily worked in sheets and small objects. Celluloid can be made perfectly transparent and colorless while bakelite is confined to the range between a clear amber and an opaque brown or black. On the other hand bakelite has the advantage in being tasteless, odorless, inert, insoluble and non-inflammable. This last quality and its high electrical resistance give bakelite its chief field of usefulness. Electricity was discovered by the Greeks, who found that amber (*electron*) when rubbed would pick up straws. This means simply that amber, like all such resinous substances, natural or artificial, is a non-conductor or dielectric and does not carry off and scatter the electricity collected on the surface by the friction. Bakelite is used in its liquid form for impregnating coils to keep the wires from shortcircuiting and in its solid form for commutators, magnetos, switch blocks, distributors, and all sorts of electrical apparatus for automobiles, telephones, wireless telegraphy, electric lighting, etc.

Bakelite, however, is only one of an indefinite number of such condensation products. As Baeyer said long ago: "It seems that all the aldehydes will, under suitable circumstances, unite with the aromatic hydrocarbons to form resins." So instead of phenol, other coal tar products such as cresol, naphthol or benzene itself may be used. The carbon links ($-CH_2-$, methylene) necessary to hook these carbon rings together may be obtained from other substances than the aldehydes, for instance from the amines, or ammonia derivatives. Three chemists, L. V. Redman, A. J. Weith and F. P. Broek, working in 1910 on the Industrial Fellowships of the late Robert Kennedy Duncan at the University of Kansas, developed a process using formin instead of formaldehyde. Formin—or, if you insist upon its full name, hexa-methylene-tetramine—is a sugar-like substance with a fish-like smell. This mixed with crystallized carbolic acid and slightly warmed melts to a golden liquid that sets on pouring into molds. It is still plastic and can be bent into any desired shape, but on further heating it becomes hard without the need of pressure. Ammonia is given off in this process instead of water which is the by-product in the case of formaldehyde. The product is similar to bakelite, exactly how similar is a question that the courts will have to decide. The inventors threatened to call it Phenyl-endeka-saligeno-saligenin, but, rightly fearing that this would interfere with its salability, they have named it "redmanol."

A phenolic condensation product closely related to bakelite and redmanol is condensite, the invention of Jonas Walter Aylesworth. Aylesworth was trained in what he referred to as "the greatest university of the world, the Edison laboratory." He entered this university at the age of nineteen at a salary of $3 a week, but Edison soon found that he had in his new boy an assistant who could stand being shut up in the laboratory working day and night as long as he could. After nine years of close association with Edison he set up a little laboratory in his own back yard to work out new plastics. He found that by acting on naphthalene—the moth-ball stuff—with chlorine he got a series of useful products called "halowaxes." The lower chlorinated products are oils, which may be used for impregnating paper or soft wood, making it noninflammable and impregnable to water. If four atoms of chlorine enter the naphthalene molecule the product is a hard wax that rings like a metal.

Condensite is anhydrous and infusible, and like its rivals finds its chief employment in the insulation parts of electrical apparatus. The records of the Edison phonograph are made of it. So are the buttons of our blue-jackets. The Government at the outbreak of the war ordered 40,000 goggles in condensite frames to protect the eyes of our gunners from the glare and acid fumes.

The various synthetics played an important part in the war. According to an ancient military pun the endurance of soldiers depends upon the strength of their soles. The new compound rubber soles were found useful in our army and the Germans attribute their success in making a little leather go a long way during the late war to the use of a new synthetic tanning material known as "neradol." There are various forms of this. Some are phenolic condensation products of formaldehyde like those we have been considering, but some use coal-tar compounds having no phenol groups, such as naphthalene sulfonic acid. These are now being made in England under such names as "paradol," "cresyntan" and "syntan." They have the advantage of the natural tannins such as bark in that they are of known strength and can be varied to suit.

This very grasping compound, formaldehyde, will attack almost anything, even molecules many times its size. Gelatinous and albuminous substances of all sorts are solidified by it. Glue, skimmed milk, blood, eggs, yeast, brewer's slops, may by this magic agent be rescued from waste and reappear in our buttons, hairpins, roofing, phonographs, shoes or shoe-polish. The French have made great use of casein hardened by formaldehyde into what is known as "galalith" (i. e., milkstone). This is harder than celluloid and non-inflammable, but has the disadvantages of being more brittle and of absorbing moisture. A mixture of casein and celluloid has something of the merits of both.

The Japanese, as we should expect, are using the juice of the soy bean, familiar as a condiment to all who patronize chop-sueys or use Worcestershire sauce. The soy glucine coagulated by formalin gives a plastic said to be better and cheaper than celluloid. Its inventor, S. Sato, of Sendai University, has named it, according to American precedent, "Satolite," and has organized a million-dollar Satolite Company at Mukojima.

The algin extracted from the Pacific kelp can be used as a rubber surrogate for waterproofing cloth. When combined with heavier alkaline bases it forms a tough and elastic substance that can be rolled into transparent sheets like celluloid or turned into buttons and knife handles.

In Australia when the war shut off the supply of tin the Government commission appointed to devise means of preserving fruits recommended the use of cardboard containers varnished with "magramite." This is a name the Australians coined for synthetic resin made from phenol and formaldehyde like bakelite. Magramite dissolved in alcohol is painted on the cardboard cans and when these are stoved the coating becomes insoluble.

Tarasoff has made a series of condensation products from phenol and formaldehyde with the addition of sulfonated oils. These are formed by the action of sulfuric acid on coconut, castor, cottonseed or mineral oils. The products of this combination are white plastics, opaque, insoluble and infusible.

Since I am here chiefly concerned with "Creative Chemistry," that is, with the art of making substances not found in nature, I have not spoken of shellac, asphaltum, rosin, ozocerite and the innumerable gums, resins and waxes, animal, mineral and vegetable, that are used either by themselves or in combination with the synthetics. What particular "dope" or "mud" is used to coat a canvas or form a telephone receiver is often hard to find out. The manufacturer finds secrecy safer than the patent office and the chemist of a rival establishment is apt to be baffled in his attempt to analyze and imitate. But we of the outside world are not concerned with this, though we are interested in the manifold applications of these new materials.

There seems to be no limit to these compounds and

every week the journals report new processes and patents. But we must not allow the new ones to crowd out the remembrance of the oldest and most famous of the synthetic plasters, hard rubber, to which a separate chapter must be devoted.

VIII

THE RACE FOR RUBBER

There is one law that regulates all animate and inanimate things. It is formulated in various ways, for instance:

Running down a hill is easy. In Latin it reads, *facilis descensus Averni.* Herbert Spencer calls it the dissolution of definite coherent heterogeneity into indefinite incoherent homogeneity. Mother Goose expresses it in the fable of Humpty Dumpty, and the business man extracts the moral as, "You can't unscramble an egg." The theologian calls it the dogma of natural depravity. The physicist calls it the second law of thermodynamics. Clausius formulates it as "The entropy of the world tends toward a maximum." It is easier to smash up than to build up. Children find that this is true of their toys; the Bolsheviki have found that it is true of a civilization. So, too, the chemist knows analysis is easier than synthesis and that creative chemistry is the highest branch of his art.

This explains why chemists discovered how to take rubber apart over sixty years before they could find out how to put it together. The first is easy. Just put some raw rubber into a retort and heat it. If you can stand the odor you will observe the caoutchouc decomposing and a benzine-like liquid distilling over.

This is called "isoprene." Any Freshman chemist could write the reaction for this operation. It is simply

$$C_{10}H_{16} \rightarrow 2C_5H_8$$
$$\text{caoutchouc} \quad\quad\quad \text{isoprene}$$

That is, one molecule of the gum splits up into two molecules of the liquid. It is just as easy to write the reaction in the reverse directions, as 2 isoprene → 1 caoutchouc, but nobody could make it go in that direction. Yet it could be done. It had been done. But the man who did it did not know how he did it and could not do it again. Professor Tilden in May, 1892, read a paper before the Birmingham Philosophical Society in which he said:

> I was surprised a few weeks ago at finding the contents of the bottles containing isoprene from turpentine entirely changed in appearance. In place of a limpid, colorless liquid the bottles contained a dense syrup in which were floating several large masses of a yellowish color. Upon examination this turned out to be India rubber.

But neither Professor Tilden nor any one else could repeat this accidental metamorphosis. It was tantalizing, for the world was willing to pay $2,000,000,000 a year for rubber and the forests of the Amazon and Congo were failing to meet the demand. A large share of these millions would have gone to any chemist who could find out how to make synthetic rubber and make it cheaply enough. With such a reward of fame and fortune the competition among chemists was intense. It took the form of an international contest in which England and Germany were neck and neck.

Courtesy of the "India Rubber World."

What goes into rubber and what is made out of it

The English, who had been beaten by the Germans in the dye business where they had the start, were determined not to lose in this. Prof. W. H. Perkin, of Manchester University, was one of the most eager, for he was inspired by a personal grudge against the Germans as well as by patriotism and scientific zeal. It was his father who had, fifty years before, discovered mauve, the first of the anilin dyes, but England could not hold the business and its rich rewards went over to Germany. So in 1909 a corps of chemists set to work under Professor Perkin in the Manchester laboratories to solve the problem of synthetic rubber. What reagent could be found that would reverse the reaction and convert the liquid isoprene into the solid rubber? It was discovered, by accident, we may say, but it should be understood that such advantageous accidents happen only to those who are working for them and know how to utilize them. In July, 1910, Dr. Matthews, who had charge of the research, set some isoprene to drying over metallic sodium, a common laboratory method of freeing a liquid from the last traces of water. In September he found that the flask was filled with a solid mass of real rubber instead of the volatile colorless liquid he had put into it.

Twenty years before the discovery would have been useless, for sodium was then a rare and costly metal, a little of it in a sealed glass tube being passed around the chemistry class once a year as a curiosity, or a tiny bit cut off and dropped in water to see what a fuss it made. But nowadays metallic sodium is cheaply produced by the aid of electricity. The difficulty lay rather in the cost of the raw material, isoprene. In industrial chemistry it is not sufficient that a thing can be made; it must be made to pay. Isoprene could be obtained from turpentine, but this was too expensive and limited in supply. It would merely mean the destruction of pine forests instead of rubber forests. Starch was finally decided upon as the best material, since this can be obtained for about a cent a pound from potatoes, corn and many other sources. Here, however, the chemist came to the end of his rope and had to call the bacteriologist to his aid. The splitting of the starch molecule is too big a job for man; only the lower organisms, the yeast plant, for example, know enough to do that. Owing perhaps to the *entente cordiale* a French biologist was called into the combination, Professor Fernbach, of the Pasteur Institute, and after eighteen months' hard work he discovered a process of fermentation by which a large amount of fusel oil can be obtained from any starchy stuff. Hitherto the aim in fermentation and distillation had been to obtain as small a proportion of fusel as possible, for fusel oil is a mixture of the heavier alcohols, all of them more poisonous and malodorous than common alcohol. But here, as has often happened in the history of industrial chemistry, the by-product turned out to be more valuable than the product. From fusel oil by the use of chlorine isoprene can be prepared, so the chain was complete.

But meanwhile the Germans had been making equal progress. In 1905 Prof. Karl Harries, of Berlin, found out the name of the caoutchouc molecule. This discovery was to the chemists what the architect's plan of a house is to the builder. They knew then what they were trying to construct and could go about their task intelligently.

Mark Twain said that he could understand something about how astronomers could measure the distance of the planets, calculate their weights and so forth, but he never could see how they could find out their names even with the largest telescopes. This is a joke in astronomy but it is not in chemistry. For when the chemist finds out the structure of a compound he gives it a name which means that. The stuff came to be called "caoutchouc," because that was the way the Spaniards of Columbus's time caught the Indian word "cahuchu." When Dr. Priestley called it "India rubber" he told merely where it came from and what it was good for. But when Harries named it "1-5-dimethyl-cyclo-octadien-1-5" any chemist could draw a picture of it and give a guess as to how it could be made. Even a person without any knowledge of chemistry can get the main point of it by merely looking at this diagram:

isoprene *turns into* caoutchouc

I have dropped the 16 H's or hydrogen atoms of the formula for simplicity's sake. They simply hook on wherever they can. You will see that the isoprene consists of a chain of four carbon atoms (represented by the C's) with an extra carbon on the side. In the transformation of this colorless liquid into soft rubber two of the double linkages break and so permit the two chains of 4 C's to unite to form one ring of eight. If you have ever played ring-around-a-rosy you will get the idea. In Chapter IV I explained that the anilin dyes are built up upon the benzene ring of six carbon atoms. The rubber ring consists of eight at least and probably more. Any substance containing that peculiar carbon chain with two double links C—C—C=C can double up—polymerize, the chemist calls it—into a rubber-like substance. So we may have many kinds of rubber, some of which may prove to be more useful than that which happens to be found in nature.

With the structural formula of Harries as a clue chemists all over the world plunged into the problem with renewed hope. The famous Bayer dye works at Elberfeld took it up and there in August, 1909, Dr. Fritz Hofmann worked out a process for the converting of pure isoprene into rubber by heat. Then in 1910 Harries happened upon the same sodium reaction as Matthews, but when he came to get it patented he found that the Englishman had beaten him to the patent office by a few weeks.

This Anglo-German rivalry came to a dramatic climax in 1912 at the great hall of the College of the City of New York when Dr. Carl Duisberg, of the Elberfeld factory, delivered an address on the latest achievements of the chemical industry before the Eighth—and the last for a long time—International Congress of Applied Chemistry. Duisberg insisted upon talking in German, although more of his auditors would have understood him in English. He laid full emphasis upon German achievements and cast doubt upon the claim of "the Englishman Tilden" to have prepared artificial rubber in the eighties. Perkin, of Manchester, confronted him with his new process for making rubber from potatoes, but Duisberg countered by proudly displaying two automobile tires made of synthetic rubber with which he had made a thousand-mile run.

The intense antagonism between the British and German chemists at this congress was felt by all present, but we did not foresee that in two years from that date they would be engaged in manufacturing poison gas to fire at one another. It was, however, realized that more was at stake than personal reputation and national prestige. Under pressure of the new demand for automobiles the price of rubber jumped from $1.25 to $3 a pound in 1910, and millions had been invested in plantations. If Professor Perkin was right when he told the congress that by his process rubber could be made for less than 25 cents a pound it meant that these plantations would go the way of the indigo plantations when the Germans succeeded in making artificial indigo. If Dr. Duisberg was right when he told the congress that synthetic rubber would "certainly appear on the market in a very short time," it meant that Germany in war or peace would become independent of Brazil in the matter of rubber as she had become independent of Chile in the matter of nitrates.

As it turned out both scientists were too sanguine. Synthetic rubber has not proved capable of displacing natural rubber by underbidding it nor even of replac-

THE RACE FOR RUBBER

ing natural rubber when this is shut out. When Germany was blockaded and the success of her armies depended on rubber, price was no object. Three Danish sailors who were caught by United States officials trying to smuggle dental rubber into Germany confessed that they had been selling it there for gas masks at $73 a pound. The German gas masks in the latter part of the war were made without rubber and were frail and leaky. They could not have withstood the new gases which American chemists were preparing on an unprecedented scale. Every scrap of old rubber in Germany was saved and worked over and over and diluted with fillers and surrogates to the limit of elasticity. Spring tires were substituted for pneumatics. So it is evident that the supply of synthetic rubber could not have been adequate or satisfactory. Neither, on the other hand, have the British made a success of the Perkin process, although they spent $200,000 on it in the first two years. But, of course, there was not the same necessity for it as in the case of Germany, for England had practically a monopoly of the world's supply of natural rubber either through owning plantations or controlling shipping. If rubber could not be manufactured profitably in Germany when the demand was imperative and price no consideration it can hardly be expected to compete with the natural under peace conditions.

The problem of synthetic rubber has then been solved scientifically but not industrially. It can be made but cannot be made to pay. The difficulty is to find a cheap enough material to start with. We can make rubber out of potatoes—but potatoes have other uses. It would require more land and more valuable land to raise the potatoes than to raise the rubber. We can get isoprene by the distillation of turpentine—but why not bleed a rubber tree as well as a pine tree? Turpentine is neither cheap nor abundant enough. Any kind of wood, sawdust for instance, can be utilized by converting the cellulose over into sugar and fermenting this to alcohol, but the process is not likely to prove profitable. Petroleum when cracked up to make gasoline gives isoprene or other double-bond compounds that go over into some form of rubber.

But the most interesting and most promising of all is the complete inorganic synthesis that dispenses with the aid of vegetation and starts with coal and lime. These heated together in the electric furnace form calcium carbide and this, as every automobilist knows, gives acetylene by contact with water. From this gas isoprene can be made and the isoprene converted into rubber by sodium, or acid or alkali or simple heating. Acetone, which is also made from acetylene, can be converted directly into rubber by fuming sulfuric acid. This seems to have been the process chiefly used by the Germans during the war. Several carbide factories were devoted to it. But the intermediate and by-products of the process, such as alcohol, acetic acid and acetone, were in as much demand for war purposes as rubber. The Germans made some rubber from pitch imported from Sweden. They also found a useful substitute in aluminum naphthenate made from Baku petroleum, for it is elastic and plastic and can be vulcanized.

So although rubber can be made in many different

ways it is not profitable to make it in any of them. We have to rely still upon the natural product, but we can greatly improve upon the way nature produces it. When the call came for more rubber for the electrical and automobile industries the first attempt to increase the supply was to put pressure upon the natives to bring in more of the latex. As a consequence the trees were bled to death and sometimes also the natives. The Belgian atrocities in the Congo shocked the civilized world and at Putumayo on the upper Amazon the same cause produced the same horrible effects. But no matter what cruelty was practiced the tropical forests could not be made to yield a sufficient increase, so the cultivation of the rubber was begun by far-sighted men in Dutch Java, Sumatra and Borneo and in British Malaya and Ceylon.

Brazil, feeling secure in the possession of a natural monopoly, made no effort to compete with these parvenus. It cost about as much to gather rubber from the Amazon forests as it did to raise it on a Malay plantation, that is, 25 cents a pound. The Brazilian Government clapped on another 25 cents export duty and spent the money lavishly. In 1911 the treasury of Para took in $2,000,000 from the rubber tax and a good share of the money was spent on a magnificent new theater at Manaos—not on setting out rubber trees. The result of this rivalry between the collector and the cultivator is shown by the fact that in the decade 1907–1917 the world's output of plantation rubber increased from 1000 to 204,000 tons, while the output of wild rubber decreased from 68,000 to 53,000. Besides this the plantation rubber is a cleaner and more even product, carefully coagulated by acetic acid instead of being smoked over a forest fire. It comes in pale yellow sheets instead of big black balls loaded with the dirt or sticks and stones that the honest Indian sometimes adds to make a bigger lump. What's better, the man who milks the rubber trees on a plantation may live at home where he can be decently looked after. The agriculturist and the chemist may do what the philanthropist and statesman could not accomplish: put an end to the cruelties involved in the international struggle for "black gold."

The United States uses three-fourths of the world's rubber output and grows none of it. What is the use of tropical possessions if we do not make use of them? The Philippines could grow all our rubber and keep a $300,000,000 business under our flag. Santo Domingo, where rubber was first discovered, is now under our supervision and could be enriched by the industry. The Guianas, where the rubber tree was first studied, might be purchased. It is chiefly for lack of a definite colonial policy that our rubber industry, by far the largest in the world, has to be dependent upon foreign sources for all its raw materials. Because the Philippines are likely to be cast off at any moment, American maufacturers are placing their plantations in the Dutch or British possessions. The Goodyear Company has secured a concession of 20,000 acres near Medan in Dutch Sumatra.

While the United States is planning to relinquish its Pacific possessions the British have more than doubled their holdings in New Guinea by the acquisition of Kaiser Wilhelm's Land, good rubber country.

THE RACE FOR RUBBER

The British Malay States in 1917 exported over $118,-000,000 worth of plantation-grown rubber and could have sold more if shipping had not been short and production restricted. Fully 90 per cent. of the cultivated rubber is now grown in British colonies or on British plantations in the Dutch East Indies. To protect this monopoly an act has been passed preventing foreigners from buying more land in the Malay Peninsula. The Japanese have acquired there 50,000 acres, on which they are growing more than a million dollars' worth of rubber a year. The British *Tropical Life* says of the American invasion: "As America is so extremely wealthy Uncle Sam can well afford to continue to buy our rubber as he has been doing instead of coming in to produce rubber to reduce his competition as a buyer in the world's market." The Malaya estates calculate to pay a dividend of 20 per cent. on the investment with rubber selling at 30 cents a pound and every two cents additional on the price brings a further 3½ per cent. dividend. The output is restricted by the Rubber Growers' Association so as to keep the price up to 50–70 cents. When the plantations first came into bearing in 1910 rubber was bringing nearly $3 a pound, and since it can be produced at less than 30 cents a pound we can imagine the profits of the early birds.

The fact that the world's rubber trade was in the control of Great Britain caused America great anxiety and financial loss in the early part of the war when the British Government, suspecting—not without reason—that some American rubber goods were getting into Germany through neutral nations, suddenly shut off our supply. This threatened to kill the fourth largest of our industries and it was only by the submission of American rubber dealers to the closest supervision and restriction by the British authorities that they were allowed to continue their business. Sir Francis Hopwood, in laying down these regulations, gave emphatic warning "that in case any manufacturer, importer or dealer came under suspicion his permits should be immediately revoked. Reinstatement will be slow and difficult. The British Government will cancel first and investigate afterward." Of course the British had a right to say under what conditions they should sell their rubber and we cannot blame them for taking such precautions to prevent its getting to their enemies, but it placed the United States in a humiliating position and if we had not been in sympathy with their side it would have aroused more resentment than it did. But it made evident the desirability of having at least part of our supply under our own control and, if possible, within our own country. Rubber is not rare in nature, for it is contained in almost every milky juice. Every country boy knows that he can get a self-feeding mucilage brush by cutting off a milkweed stalk. The only native source so far utilized is the guayule, which grows wild on the deserts of the Mexican and the American border. The plant was discovered in 1852 by Dr. J. M. Bigelow near Escondido Creek, Texas. Professor Asa Gray described it and named it Parthenium argentatum, or the silver Pallas. When chopped up and macerated guayule gives a satisfactory quality of caoutchouc in profitable amounts. In 1911 seven thousand tons of guayule were imported from Mexico; in 1917 only seventeen hundred tons. Why this falling off? Because the eager exploiters had killed the goose that laid the golden egg, or in plain language, pulled up the plant by the roots. Now guayule is being cultivated and is reaped instead of being uprooted. Experiments at the Tucson laboratory have recently removed the difficulty of getting the seed to germinate under cultivation. This seems the most promising of the home-grown plants and, until artificial rubber can be made profitable, gives us the only chance of being in part independent of oversea supply.

There are various other gums found in nature that can for some purposes be substituted for caoutchouc. Gutta percha, for instance, is pliable and tough though not very elastic. It becomes plastic by heat so it can be molded, but unlike rubber it cannot be hardened by heating with sulfur. A lump of gutta percha was brought from Java in 1766 and placed in a British museum, where it lay for nearly a hundred years before it occurred to anybody to do anything with it except to look at it. But a German electrician, Siemens, discovered in 1847 that gutta percha was valuable for insulating telegraph lines and it found extensive employment in submarine cables as well as for golf balls, and the like.

Balata, which is found in the forests of the Guianas, is between gutta percha and rubber, not so good for insulation but useful for shoe soles and machine belts. The bark of the tree is so thick that the latex does not run off like caoutchouc when the bark is cut. So the bark has to be cut off and squeezed in hand presses.

Formerly this meant cutting down the tree, but now alternate strips of the bark are cut off and squeezed so the tree continues to live.

When Columbus discovered Santo Domingo he found the natives playing with balls made from the gum of the caoutchouc tree. The soldiers of Pizarro, when they conquered Inca-Land, adopted the Peruvian custom of smearing caoutchouc over their coats to keep out the rain. A French scientist, M. de la Condamine, who went to South America to measure the earth, came back in 1745 with some specimens of caoutchouc from Para as well as quinine from Peru. The vessel on which he returned, the brig *Minerva*, had a narrow escape from capture by an English cruiser, for Great Britain was jealous of any trespassing on her American sphere of influence. The Old World need not have waited for the discovery of the New, for the rubber tree grows wild in Annam as well as Brazil, but none of the Asiatics seems to have discovered any of the many uses of the juice that exudes from breaks in the bark.

The first practical use that was made of it gave it the name that has stuck to it in English ever since. Magellan announced in 1772 that it was good to remove pencil marks. A lump of it was sent over from France to Priestley, the clergyman chemist who discovered oxygen and was mobbed out of Manchester for being a republican and took refuge in Pennsylvania. He cut the lump into little cubes and gave them to his friends to eradicate their mistakes in writing or figuring. Then they asked him what the queer things were and he said that they were "India rubbers."

The Peruvian natives had used caoutchouc for water

proof clothing, shoes, bottles and syringes, but Europe was slow to take it up, for the stuff was too sticky and smelled too bad in hot weather to become fashionable in fastidious circles. In 1825 Mackintosh made his name immortal by putting a layer of rubber between two cloths.

A German chemist, Ludersdorf, discovered in 1832 that the gum could be hardened by treating it with sulfur dissolved in turpentine. But it was left to a Yankee inventor, Charles Goodyear, of Connecticut, to work out a practical solution of the problem. A friend of his, Hayward, told him that it had been revealed to him in a dream that sulfur would harden rubber, but unfortunately the angel or defunct chemist who inspired the vision failed to reveal the details of the process. So Hayward sold out his dream to Goodyear, who spent all his own money and all he could borrow from his friends trying to convert it into a reality. He worked for ten years on the problem before the "lucky accident" came to him. One day in 1839 he happened to drop on the hot stove of the kitchen that he used as a laboratory a mixture of caoutchouc and sulfur. To his surprise he saw the two substances fuse together into something new. Instead of the soft, tacky gum and the yellow, brittle brimstone he had the tough, stable, elastic solid that has done so much since to make our footing and wheeling safe, swift and noiseless. The gumshoes or galoshes that he was then enabled to make still go by the name of "rubbers" in this country, although we do not use them for pencil erasers.

Goodyear found that he could vary this "vulcanized

rubber" at will. By adding a little more sulfur he got a hard substance which, however, could be softened by heat so as to be molded into any form wanted. Out of this "hard rubber" "vulcanite" or "ebonite" were made combs, hairpins, penholders and the like, and it has not yet been superseded for some purposes by any of its recent rivals, the synthetic resins.

The new form of rubber made by the Germans, methyl rubber, is said to be a superior substitute for the hard variety but not satisfactory for the soft. The electrical resistance of the synthetic product is 20 per cent. higher than the natural, so it is excellent for insulation, but it is inferior in elasticity. In the latter part of the war the methyl rubber was manufactured at the rate of 165 tons a month.

The first pneumatic tires, known then as "patent aerial wheels," were invented by Robert William Thomson of London in 1846. On the following year a carriage equipped with them was seen in the streets of New York City. But the pneumatic tire did not come into use until after 1888, when an Irish horse-doctor, John Boyd Dunlop, of Belfast, tied a rubber tube around the wheels of his little son's velocipede. Within seven years after that a $25,000,000 corporation was manufacturing Dunlop tires. Later America took the lead in this business. In 1913 the United States exported $3,000,000 worth of tires and tubes. In 1917 the American exports rose to $13,000,000, not counting what went to the Allies. The number of pneumatic tires sold in 1917 is estimated at 18,000,000, which at an average cost of $25 would amount to $450,000,000.

THE RACE FOR RUBBER

No matter how much synthetic rubber may be manufacturer or how many rubber trees are set out there is no danger of glutting the market, for as the price falls the uses of rubber become more numerous. One can think of a thousand ways in which rubber could be used if it were only cheap enough. In the form of pads and springs and tires it would do much to render traffic noiseless. Even the elevated railroad and the subway might be opened to conversation, and the city made habitable for mild voiced and gentle folk. It would make one's step sure, noiseless and springy, whether it was used individualistically as rubber heels or collectivistically as carpeting and paving. In roofing and siding and paint it would make our buildings warmer and more durable. It would reduce the cost and permit the extension of electrical appliances of almost all kinds. In short, there is hardly any other material whose abundance would contribute more to our comfort and convenience. Noise is an automatic alarm indicating lost motion and wasted energy. Silence is economy and resiliency is superior to resistance. A gumshoe outlasts a hobnailed sole and a rubber tube full of air is better than a steel tire.

IX

THE RIVAL SUGARS

The ancient Greeks, being an inquisitive and acquisitive people, were fond of collecting tales of strange lands. They did not care much whether the stories were true or not so long as they were interesting. Among the marvels that the Greeks heard from the Far East two of the strangest were that in India there were plants that bore wool without sheep and reeds that bore honey without bees. These incredible tales turned out to be true and in the course of time Europe began to get a little calico from Calicut and a kind of edible gravel that the Arabs who brought it called "sukkar." But of course only kings and queens could afford to dress in calico and have sugar prescribed for them when they were sick.

Fortunately, however, in the course of time the Arabs invaded Spain and forced upon the unwilling inhabitants of Europe such instrumentalities of higher civilization as arithmetic and algebra, soap and sugar. Later the Spaniards by an act of equally unwarranted and beneficent aggression carried the sugar cane to the Caribbean, where it thrived amazingly. The West Indies then became a rival of the East Indies as a treasure-house of tropical wealth and for several centuries the Spanish, Portuguese, Dutch, English, Danes and French fought like wildcats to gain possession of this little nest of islands and the routes leading thereunto.

The English finally overcame all these enemies, whether they fought her singly or combined. Great Britain became mistress of the seas and took such Caribbean lands as she wanted. But in the end her continental foes came out ahead, for they rendered her victory valueless. They were defeated in geography but they won in chemistry. Canning boasted that "the New World had been called into existence to redress the balance of the Old." Napoleon might have boasted that he had called in the sugar beet to balance the sugar cane. France was then, as Germany was a century later, threatening to dominate the world. England, then as in the Great War, shut off from the seas the shipping of the aggressive power. France then, like Germany later, felt most keenly the lack of tropical products, chief among which, then but not in the recent crisis, was sugar. The cause of this vital change is that in 1747 Marggraf, a Berlin chemist, discovered that it was possible to extract sugar from beets. There was only a little sugar in the beet root then, some six per cent., and what he got out was dirty and bitter. One of his pupils in 1801 set up a beet sugar factory near Breslau under the patronage of the King of Prussia, but the industry was not a success until Napoleon took it up and in 1810 offered a prize of a million francs for a practical process. How the French did make fun of him for this crazy notion! In a comic paper of that day you will find a cartoon of Napoleon in the nursery beside the cradle of his son and heir, the King of Rome—known to the readers of Rostand as l'Aiglon. The Emperor is squeezing the juice of a beet into his coffee and the nurse has put a beet into the mouth of the infant King, saying: "Suck, dear, suck. Your father says it's sugar."

In like manner did the wits ridicule Franklin for fooling with electricity, Rumford for trying to improve chimneys, Parmentier for thinking potatoes were fit to eat, and Jefferson for believing that something might be made of the country west of the Mississippi. In all ages ridicule has been the chief weapon of conservatism. If you want to know what line human progress will take in the future read the funny papers of today and see what they are fighting. The satire of every century from Aristophanes to the latest vaudeville has been directed against those who are trying to make the world wiser or better, against the teacher and the preacher, the scientist and the reformer.

In spite of the ridicule showered upon it the despised beet year by year gained in sweetness of heart. The percentage of sugar rose from six to eighteen and by improved methods of extraction became finally as pure and palatable as the sugar of the cane. An acre of German beets produces more sugar than an acre of Louisiana cane. Continental Europe waxed wealthy while the British West Indies sank into decay. As the beets of Europe became sweeter the population of the islands became blacker. Before the war England was paying out $125,000,000 for sugar, and more than two-thirds of this money was going to Germany and Austria-Hungary. Fostered by scientific study, protected by tariff duties, and stimulated by export bounties, the beet sugar industry became one of the financial forces of the world. The English at home, especially the marmalade-makers, at first rejoiced at the idea of getting sugar for less than cost at the expense of her continental rivals. But the suffering colonies took another

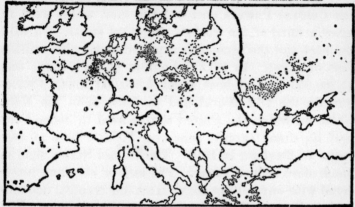

MAP SHOWING LOCATION OF EUROPEAN BEET SUGAR FACTORIES—ALSO BATTLE LINES AT CLOSE OF 1916
ESTIMATED THAT ONE-THIRD OF WORLD'S PRODUCTION BEFORE THE WAR WAS PRODUCED WITHIN BATTLE LINES

Courtesy American Sugar Refining Co.

view of the situation. In 1888 a conference of the powers called at London agreed to stop competing by the pernicious practice of export bounties, but France and the United States refused to enter, so the agreement fell through. Another conference ten years later likewise failed, but when the parvenu beet sugar ventured to invade the historic home of the cane the limit of toleration had been reached. The Council of India put on countervailing duties to protect their home-grown cane from the bounty-fed beet. This forced the calling of a convention at Brussels in 1903 "to equalize the conditions of competition between beet sugar and cane sugar of the various countries," at which the powers agreed to a mutual suppression of bounties. Beet sugar then divided the world's market equally with cane sugar and the two rivals stayed substantially neck and neck until the Great War came. This shut out from England the product of Germany, Austria-Hungary, Belgium, northern France and Russia and took the farmers from their fields. The battle lines of the Central Powers enclosed the land which used to grow a third of the world's supply of sugar. In 1913 the beet and the cane each supplied about nine million tons of sugar. In 1917 the output of cane sugar was 11,200,000 and of beet sugar 5,300,000 tons. Consequently the Old World had to draw upon the New. Cuba, on which the United States used to depend for half its sugar supply, sent over 700,000 tons of raw sugar to England in 1916. The United States sent as much more refined sugar. The lack of shipping interfered with our getting sugar from our tropical dependencies, Hawaii, Porto Rico and the Philippines. The homegrown beets give us only a fifth and the cane of Louisiana and Texas only a fifteenth of the sugar we need. As a result we were obliged to file a claim in advance to get a pound of sugar from the corner grocery and then we were apt to be put off with rock candy, muscovado or honey. Lemon drops proved useful for Russian tea and the "long sweetening" of our forefathers came again into vogue in the form of various syrups. The United States was accustomed to consume almost a fifth of all the sugar produced in the world—and then we could not get it.

The shortage made us realize how dependent we have become upon sugar. Yet it was, as we have seen, practically unknown to the ancients and only within the present generation has it become an essential factor in

© Underwood & Underwood

FOREST RUBBER

Compare this tropical tangle and gnarled trunk with the straight tree and cleared ground of the plantation. At the foot of the trunk are cups collecting rubber juice.

© Publishers' Photo Service

PLANTATION RUBBER

This spiral cut draws off the milk as completely and quickly as possible without harming the tree. The man is pulling off a strip of coagulated rubber that clogs it.

© Underwood & Underwood

IN MAKING GARDEN HOSE THE RUBBER IS FORMED INTO A TUBE BY THE MACHINE ON THE RIGHT AND COILED ON THE TABLE TO THE LEFT

our diet. As soon as the chemist made it possible to produce sugar at a reasonable price all nations began

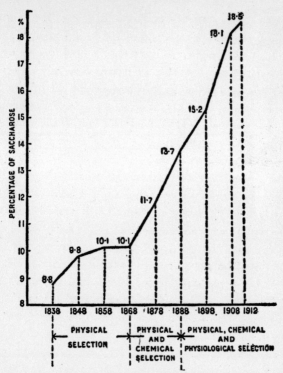

How the sugar beet has gained enormously in sugar content under chemical control

to buy it in proportion to their means. Americans, as the wealthiest people in the world, ate the most, ninety pounds a year on the average for every man, woman and child. In other words we ate our weight of sugar every year. The English consumed nearly as much as the Americans; the French and Germans about half as much; the Balkan peoples less than ten pounds per annum; and the African savages none.

Pure white sugar is the first and greatest contribution of chemistry to the world's dietary. It is unique in being a single definite chemical compound, sucrose, $C_{12}H_{22}O_{11}$. All natural nutriments are more or less complex mixtures. Many of them, like wheat or milk or fruit, contain in various proportions all of the three factors of foods, the fats, the proteids and the carbohydrates, as well as water and the minerals and other ingredients necessary to life. But sugar is a simple substance, like water or salt, and like them is incapable of sustaining life alone, although unlike them it is nutritious. In fact, except the fats there is no more nutritious food than sugar, pound for pound, for it contains no water and no waste. It is therefore the quickest and usually the cheapest means of supplying bodily energy. But as may be seen from its formula as given above it contains only three elements, carbon, hydrogen and oxygen, and omits nitrogen and other elements necessary to the body. An engine requires not only coal but also lubricating oil, water and bits of steel and brass to keep it in repair. But as a source of the energy needed in our strenuous life sugar has no equal and only one rival, alcohol. Alcohol is the offspring of sugar, a degenerate descendant that retains but few of the good qualities of its sire and has acquired some evil traits of its own. Alcohol, like sugar, may serve to furnish the energy of a steam engine or a human body. Used as a fuel alcohol has certain advantages, but used as a food it has the disqualification of deranging the bodily mechanism. Even a little alcohol will impair the accuracy and speed of thought and action, while a large quantity, as we all know from observation if not experience, will produce temporary incapacitation.

When man feeds on sugar he splits it up by the aid of air into water and carbon dioxide in this fashion:

$$\underset{\text{cane sugar}}{C_{12}H_{22}O_{11}} + \underset{\text{oxygen}}{12O_2} \rightarrow \underset{\text{water}}{11H_2O} + \underset{\text{carbon dioxide}}{12CO_2}$$

When sugar is burned the reaction is just the same.

But when the yeast plant feeds on sugar it carries the process only part way and instead of water the product is alcohol, a very different thing, so they say who have tried both as beverages. The yeast or fermentation reaction is this:

$$\underset{\text{cane sugar}}{C_{12}H_{22}O_{11}} + \underset{\text{water}}{H_2O} \rightarrow \underset{\text{alcohol}}{4C_2H_6O} + \underset{\text{carbon dioxide}}{4CO_2}$$

Alcohol then is the first product of the decomposition of sugar, a dangerous half-way house. The twin product, carbon dioxide or carbonic acid, is a gas of slightly sour taste which gives an attractive tang and effervescence to the beer, wine, cider or champagne. That is to say, one of these twins is a pestilential fellow and the other is decidedly agreeable. Yet for several thousand years mankind took to the first and let the second for the most part escape into the air. But when the chemist appeared on the scene he discovered a way of separating the two and bottling the harmless one for those who prefer it. An increasing number of people were found to prefer it, so the American soda-water fountain is gradually driving Demon Rum out of the civilized world. The brewer nowadays caters to two classes of customers. He bottles up the beer with the alcohol and a little carbonic acid in it for the saloon and he catches the rest of the carbonic acid that he used to waste and sells it to the drug stores for soda-water or uses it to charge some non-alcoholic beer of his own.

This catering to rival trades is not an uncommon thing with the chemist. As we have seen, the synthetic perfumes are used to improve the natural perfumes. Cottonseed is separated into oil and meal; the oil going to make margarin and the meal going to feed the cows that produce butter. Some people have been drinking coffee, although they do not like the taste of it, because they want the stimulating effect of its alkaloid, caffein. Other people liked the warmth and flavor of coffee but find that caffein does not agree with them. Formerly one had to take the coffee whole or let it alone. Now one can have his choice, for the caffein is extracted for use in certain popular cold drinks and the rest of the bean sold as caffein-free coffee.

Most of the "soft drinks" that are now gradually displacing the hard ones consist of sugar, water and carbonic acid, with various flavors, chiefly the esters of the fatty and aromatic acids, such as I described in a previous chapter. These are still usually made from fruits and spices and in some cases the law or public opinion requires this, but eventually, I presume, the synthetic flavors will displace the natural and then we shall get rid of such extraneous and indigestible matter as seeds, skins and bark. Suppose the world had always been used to synthetic and hence seedless figs, strawberries and blackberries. Suppose then some manufacturer of fig paste or strawberry jam should put

THE RIVAL SUGARS

in ten per cent. of little round hard wooden nodules, just the sort to get stuck between the teeth or caught in the vermiform appendix. How long would it be before he was sent to jail for adulterating food? But neither jail nor boycott has any reformatory effect on Nature.

Nature is quite human in that respect. But you can reform Nature as you can human beings by looking out for heredity and culture. In this way Mother Nature has been quite cured of her bad habit of putting seeds in bananas and oranges. Figs she still persists in adulterating with particles of cellulose as nutritious as sawdust. But we can circumvent the old lady at this. I got on Christmas a package of figs from California without a seed in them. Somebody had taken out all the seeds—it must have been a big job—and then put the figs together again as natural looking as life and very much better tasting.

Sugar and alcohol are both found in Nature; sugar in the ripe fruit, alcohol when it begins to decay. But it was the chemist who discovered how to extract them. He first worked with alcohol and unfortunately succeeded.

Previous to the invention of the still by the Arabian chemists man could not get drunk as quickly as he wanted to because his liquors were limited to what the yeast plant could stand without intoxication. When the alcoholic content of wine or beer rose to seventeen per cent. at the most the process of fermentation stopped because the yeast plants got drunk and quit "working." That meant that a man confined to ordinary wine or beer had to drink ten or twenty quarts of water to get one quart of the stuff he was after, and he had no liking for water.

So the chemist helped him out of this difficulty and got him into worse trouble by distilling the wine. The more volatile part that came over first contained the flavor and most of the alcohol. In this way he could get liquors like brandy and whisky, rum and gin, containing from thirty to eighty per cent. of alcohol. This was the origin of the modern liquor problem. The wine of the ancients was strong enough to knock out Noah and put the companions of Socrates under the table, but it was not until distilled liquors came in that alcoholism became chronic, epidemic and ruinous to whole populations.

But the chemist later tried to undo the ruin he had quite inadvertently wrought by introducing alcohol into the world. One of his most successful measures was the production of cheap and pure sugar which, as we have seen, has become a large factor in the dietary of civilized countries. As a country sobers up it takes to sugar as a "self-starter" to provide the energy needed for the strenuous life. A five o'clock candy is a better restorative than a five o'clock highball or even a five o'clock tea, for it is a true nutrient instead of a mere stimulant. It is a matter of common observation that those who like sweets usually do not like alcohol. Women, for instance, are apt to eat candy but do not commonly take to alcoholic beverages. Look around you at a banquet table and you will generally find that those who turn down their wine glasses generally take two lumps in their demi-tasses. We often hear it said that whenever a candy store opens up a saloon in the same block closes up. Our grandmothers used to warn their daughters: "Don't marry a man who does not want sugar in his tea. He is likely to take to drink." So, young man, when next you give a box of candy to your best girl and she offers you some, don't decline it. Eat it and pretend to like it, at least, for it is quite possible that she looked into a physiology and is trying you out. You never can tell what girls are up to.

In the army and navy ration the same change has taken place as in the popular dietary. The ration of rum has been mostly replaced by an equivalent amount of candy or marmalade. Instead of the tippling trooper of former days we have "the chocolate soldier." No previous war in history has been fought so largely on sugar and so little on alcohol as the last one. When the war reduced the supply and increased the demand we all felt the sugar famine and it became a mark of patriotism to refuse candy and to drink coffee unsweetened. This, however, is not, as some think, the mere curtailment of a superfluous or harmful luxury, the sacrifice of a pleasant sensation. It is a real deprivation and a serious loss to national nutrition. For there is no reason to think the constantly rising curve of sugar consumption has yet reached its maximum or optimum. Individuals overeat, but not the population as a whole. According to experiments of the Department of Agriculture men doing heavy labor may add three-quarters of a pound of sugar to their daily diet without any deleterious effects. This is at the rate of 275 pounds a year, which is three times the average consumption of England and America. But the Department does not state how much a girl doing nothing ought to eat between meals.

Of the 2500 to 3500 calories of energy required to keep a man going for a day the best source of supply is the carbohydrates, that is, the sugars and starches. The fats are more concentrated but are more expensive and less easily assimilable. The proteins are also more expensive and their decomposition products are more apt to clog up the system. Common sugar is almost an ideal food. Cheap, clean, white, portable, imperishable, unadulterated, pleasant-tasting, germ-free, highly nutritious, completely soluble, altogether digestible, easily assimilable, requires no cooking and leaves no residue. Its only fault is its perfection. It is so pure that a man cannot live on it. Four square lumps give one hundred calories of energy. But twenty-five or thirty-five times that amount would not constitue a day's ration, in fact one would ultimately starve on such fare. It would be like supplying an army with an abundance of powder but neglecting to provide any bullets, clothing or food. To make sugar the sole food is impossible. To make it the main food is unwise. It is quite proper for man to separate out the distinct ingredients of natural products—to extract the butter from the milk, the casein from the cheese, the sugar from the cane—but he must not forget to combine them again at each meal with the other essential foodstuffs in their proper proportion.

Sugar is not a synthetic product and the business of the chemist has been merely to extract and purify it. But this is not so simple as it seems and every sugar factory has had to have its chemist. He has analyzed every mother beet for a hundred years. He has watched every step of the process from the cane to the crystal lest the sucrose should invert to the less sweet and non-crystallizable glucose. He has tested with polarized light every shipment of sugar that has passed through the custom house, much to the mystification of congressmen who have often wondered at the money and argumentation expended in a tariff discussion over the question of the precise angle of rotation of the plane of vibration of infinitesimal waves in a hypothetical ether.

The reason for this painstaking is that there are dozens of different sugars, so much alike that they are difficult to separate. They are all composed of the same three elements, C, H and O, and often in the same proportion. Sometimes two sugars differ only in that one has a right-handed and the other a left-handed twist to its molecule. They bear the same resemblance to one another as the two gloves of a pair. Cane sugar and beet sugar are when completely purified the same substance, that is, sucrose, $C_{12}H_{22}O_{11}$. The brown and straw-colored sugars, which our forefathers used and which we took to using during the war, are essentially the same but have not been so completely freed from moisture and the coloring and flavoring matter of the cane juice. Maple sugar is mostly sucrose. So partly is honey. Candies are made chiefly of sucrose with the addition of glucose, gums or starch, to give them the necessary consistency and of such colors and flavors, natural or synthetic, as may be desired. Practically all candy, even the cheapest, is nowadays free from deleterious ingredients in the manufacture, though it is liable to become contaminated in the handling. In fact sugar is about the only food that is never adulterated. It would be hard to find anything cheaper to add to it that would not be easily detected. "Sanding the sugar," the crime of which grocers are generally accused, is the one they are least likely to be guilty of.

Besides the big family of sugars which are all more or less sweet, similar in structure and about equally nutritious, there are, very curiously, other chemical compounds of altogether different composition which taste like sugar but are not nutritious at all. One of these is a coal-tar derivative, discovered accidentally by an American student of chemistry, Ira Remsen, afterward president of Johns Hopkins University, and named by him "saccharin." This has the composition $C_6H_4COSO_2NH$, and as you may observe from the symbol it contains sulfur (S) and nitrogen (N) and the benzene ring (C_6H_4) that are not found in any of the sugars. It is several hundred times sweeter than sugar, though it has also a slightly bitter aftertaste. A minute quantity of it can therefore take the place of a large amount of sugar in syrups, candies and preserves, so because it lends itself readily to deception its use in food has been prohibited in the United States and other countries. But during the war, on account of the shortage of sugar, it came again into use. The European governments encouraged what they formerly tried to prevent, and it became customary in Germany or Italy to carry about a package of saccharin tablets in the pocket and drop one or two into the tea or coffee. Such reversals of ad-

THE RIVAL SUGARS

ministrative attitude are not uncommon. When the use of hops in beer was new it was prohibited by British law. But hops became customary nevertheless and now the law requires hops to be used in beer. When workingmen first wanted to form unions, laws were passed to prevent them. But now, in Australia for instance, the laws require workingmen to form unions. Governments naturally tend to a conservative reaction against anything new.

It is amusing to turn back to the pure food agitation of ten years ago and read the sensational articles in the newspapers about the poisonous nature of this dangerous drug, saccharin, in view of the fact that it is being used by millions of people in Europe in amounts greater than once seemed to upset the tender stomachs of the Washington "poison squads." But saccharin does not appear to be responsible for any fatalities yet, though people are said to be heartily sick of it. And well they may be, for it is not a substitute for sugar except to the sense of taste. Glucose may correctly be called a substitute for sucrose as margarin for butter, since they not only taste much the same but have about the same food value. But to serve saccharin in the place of sugar is like giving a rubber bone to a dog. It is reported from Europe that the constant use of saccharin gives one eventually a distaste for all sweets. This is quite likely, although it means the reversal within a few years of prehistoric food habits. Mankind has always associated sweetness with food value, for there are few sweet things found in nature except the sugars. We think we eat sugar because it is sweet. But we do not. We eat it because it is good for us.

The reason it tastes sweet to us is because it is good for us. So man makes a virtue out of necessity, a pleasure out of duty, which is the essence of ethics.

In the ancient days of Ind the great Raja Trishanku possessed an earthly paradise that had been constructed for his delectation by a magician. Therein grew all manner of beautiful flowers, savory herbs and delicious fruits such as had never been known before outside heaven. Of them all the Raja and his harems liked none better than the reed from which they could suck honey. But Indra, being a jealous god, was wroth when he looked down and beheld mere mortals enjoying such delights. So he willed the destruction of the enchanted garden. With drought and tempest it was devastated, with fire and hail, until not a leaf was left of its luxuriant vegetation and the ground was bare as a threshing floor. But the roots of the sugar cane are not destroyed though the stalk be cut down; so when men ventured to enter the desert where once had been this garden of Eden, they found the cane had grown up again and they carried away cuttings of it and cultivated it in their gardens. Thus it happened that the nectar of the gods descended first to monarchs and their favorites, then was spread among the people and carried abroad to other lands until now any child with a penny in his hand may buy of the best of it. So it has been with many things. So may it be with all things.

X

WHAT COMES FROM CORN

The discovery of America dowered mankind with a world of new flora. The early explorers in their haste to gather up gold paid little attention to the more valuable products of field and forest, but in the course of centuries their usefulness has become universally recognized. The potato and tomato, which Europe at first considered as unfit for food or even as poisonous, have now become indispensable among all classes. New World drugs like quinine and cocaine have been adopted into every pharmacopeia. Cocoa is proving a rival of tea and coffee, and even the banana has made its appearance in European markets. Tobacco and chicle occupy the nostrils and jaws of a large part of the human race. Maize and rubber are become the common property of mankind, but still may be called American. The United States alone raises four-fifths of the corn and uses three-fourths of the caoutchouc of the world.

All flesh is grass. This may be taken in a dietary as well as a metaphorical sense. The graminaceae provide the greater part of the sustenance of man and beast; hay and cereals, wheat, oats, rye, barley, rice, sugar cane, sorghum and corn. From an American viewpoint the greatest of these, physically and financially, is corn. The corn crop of the United States for 1917, amounting to 3,159,000,000 bushels, brought in more money than the wheat, cotton, potato and rye crops all together.

When Columbus reached the West Indies he found the savages playing with rubber balls, smoking incense sticks of tobacco and eating cakes made of a new grain that they called *mahiz*. When Pizarro invaded Peru he found this same cereal used by the natives not only for food but also for making alcoholic liquor, in spite of the efforts of the Incas to enforce prohibition. When the Pilgrim Fathers penetrated into the woods back of Plymouth Harbor they discovered a cache of Indian corn. So throughout the three Americas, from Canada to Peru, corn was king and it has proved worthy to rank with the rival cereals of other continents, the wheat of Europe and the rice of Asia. But food habits are hard to change and for the most part the people of the Old World are still ignorant of the delights of hasty pudding and Indian pudding, of hoe-cake and hominy, of sweet corn and popcorn. I remember thirty years ago seeing on a London stand a heap of dejected popcorn balls labeled "Novel American Confection. Please Try One." But nobody complied with this pitiful appeal but me and I was sorry that I did. Americans used to respond with a shipload of corn whenever an appeal came from famine sufferers in Armenia, Russia, Ireland, India or Austria, but their generosity was chilled when they found that their gift was resented as an insult or as an attempt to poison the impoverished population, who declared that they would rather die than eat it—and some of them did. Our Department of Agriculture sent maize missionaries to Europe with farmers and millers as educators and expert cooks to serve free flapjacks and pones, but the propaganda made little impression and today Americans are urged to eat more of their own corn because the famished families of the war-stricken region will not touch it. Just so the beggars of Munich revolted at potato soup when the pioneer of American food chemists, Rumford, attempted to introduce this transatlantic dish.

But here we are not so much concerned with corn foods as we are with its manufactured products. If you split a kernel in two you will find that it consists of three parts: a hard and horny hull on the outside, a small oily and nitrogenous germ at the point, and a white body consisting mostly of starch. Each of these is worked up into various products, as may be seen from the accompanying table. The hull forms bran and may be mixed with the gluten as a cattle food. The corn steeped for several days with sulfurous acid is disintegrated and on being ground the germs are floated off, the gluten or nitrogenous portion washed out, the starch grains settled down and the residue pressed together as oil cake fodder. The refined oil from the germ is marketed as a table or cooking oil under the name of "Mazola" and comes into competition with olive, peanut and cottonseed oil in the making of vegetable substitutes for lard and butter. Inferior grades may be used for soaps or for glycerin and perhaps nitroglycerin. A bushel of corn yields a pound or more of oil. From the corn germ also is extracted a gum called "paragol" that forms an acceptable substitute for rubber in certain uses. The "red rubber" sponges and the eraser tips to pencils may be made of it and it can contribute some twenty per cent. to the synthetic soles of shoes.

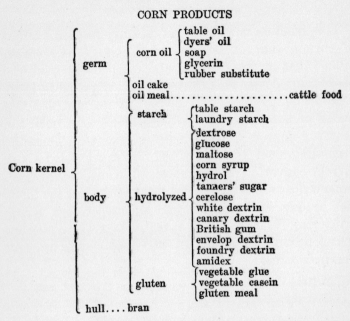

CORN PRODUCTS

Corn kernel
- germ
  - corn oil
    - table oil
    - dyers' oil
    - soap
    - glycerin
    - rubber substitute
  - oil cake
  - oil meal.....................cattle food
- body
  - starch
    - table starch
    - laundry starch
  - hydrolyzed
    - dextrose
    - glucose
    - maltose
    - corn syrup
    - hydrol
    - tanners' sugar
    - cerelose
    - white dextrin
    - canary dextrin
    - British gum
    - envelop dextrin
    - foundry dextrin
    - amidex
  - gluten
    - vegetable glue
    - vegetable casein
    - gluten meal
- hull....bran

Starch, which constitutes fifty-five per cent. of the corn kernel, can be converted into a variety of products for dietary and industrial uses. As found in corn, potatoes or any other vegetables starch consists of small, round, white, hard grains, tasteless, and insoluble in cold water. But hot water converts it into a soluble, sticky form which may serve for starching clothes or making cornstarch pudding. Carrying the process further with the aid of a little acid or other catalyst it takes up water and goes over into a sugar, dextrose,

WHAT COMES FROM CORN

commonly called "glucose." Expressed in chemical shorthand this reaction is

$$\underset{\text{starch}}{C_6H_{10}O_5} + \underset{\text{water}}{H_2O} \rightarrow \underset{\text{dextrose}}{C_6H_{12}O_6}$$

This reaction is carried out on forty million bushels of corn a year in the United States. The "starch milk," that is, the starch grains washed out from the disintegrated corn kernel by water, is digested in large pressure tanks under fifty pounds of steam with a few tenths of one per cent. of hydrochloric acid until the required degree of conversion is reached. Then the remaining acid is neutralized by caustic soda and thereby converted into common salt, which in this small amount does not interfere but rather enhances the taste. The product is the commercial glucose or corn syrup, which may if desired be evaporated to a white powder. It is a mixture of three derivatives of starch in about this proportion:

Maltose ....45 per cent.
Dextrose ....20 per cent.
Dextrin ....35 per cent.

There are also present three- or four-tenths of one per cent. salt and as much of the corn protein and a variable amount of water. It will be noticed that the glucose (dextrose), which gives name to the whole, is the least of the three ingredients.

Maltose, or malt sugar, has the same composition as cane sugar ($C_{12}H_{22}O_{11}$), but is not nearly so sweet. Dextrin, or starch paste, is not sweet at all. Dextrose or glucose is otherwise known as grape sugar, for it is commonly found in grapes and other ripe fruits. It forms half of honey and it is one of the two products into which cane sugar splits up when we take it into the mouth. It is not so sweet as cane sugar and cannot be so readily crystallized, which, however, is not altogether a disadvantage.

The process of changing starch into dextrose that takes place in the great steam kettles of the glucose factory is essentially the same as that which takes place in the ripening of fruit and in the digestion of starch. A large part of our nutriment, therefore, consists of glucose either eaten as such in ripe fruits or produced in the mouth or stomach by the decomposition of the starch of unripe fruit, vegetables and cereals. Glucose may be regarded as a predigested food. In spite of this well-known fact we still sometimes read "poor food" articles in which glucose is denounced as a dangerous adulterant and even classed as a poison.

The other ingredients of commercial glucose, the maltose and dextrin, have of course the same food value as the dextrose, since they are made over into dextrose in the process of digestion. Whether the glucose syrup is fit to eat depends, like anything else, on how it is made. If, as was formerly sometimes the case, sulfuric acid was used to effect the conversion of the starch or sulfurous acid to bleach the glucose and these acids were not altogether eliminated, the product might be unwholesome or worse. Some years ago in England there was a mysterious epidemic of arsenical poisoning among beer drinkers. On tracing it back it was found that the beer had been made from glucose which had been made from sulfuric acid which had been made from sulfur which had been made from a batch of iron pyrites which contained a little arsenic. The replacement of sulfuric acid by hydrochloric has done away with that danger and the glucose now produced is pure.

The old recipe for home-made candy called for the addition of a little vinegar to the sugar syrup to prevent "graining." The purpose of the acid was of course to invert part of the cane sugar to glucose so as to keep it from crystallizing out again. The professional candy-maker now uses the corn glucose for that purpose, so if we accuse him of "adulteration" on that ground we must levy the same accusation against our grandmothers. The introduction of glucose into candy manufacture has not injured but greatly increased the sale of sugar for the same purpose. This is not an uncommon effect of scientific progress, for as we have observed, the introduction of synthetic perfumes has stimulated the production of odoriferous flowers and the price of butter has gone up with the introduction of margarin. So, too, there are more weavers employed and they get higher wages than in the days when they smashed up the first weaving machines, and the same is true of printers and typesetting machines. The popular animosity displayed toward any new achievement of applied science is never justified, for it benefits not only the world as a whole but usually even those interests with which it seems at first to conflict.

The chemist is an economizer. It is his special business to hunt up waste products and make them useful. He was, for instance, worried over the waste of the cores and skins and scraps that were being thrown away when apples were put up. Apple pulp contains pectin, which is what makes jelly jell, and berries and fruits that are short of it will refuse to "jell." But using these for their flavor he adds apple pulp for pectin and glucose for smoothness and sugar for sweetness and, if necessary, synthetic dyes for color, he is able to put on the market a variety of jellies, jams and marmalades at very low price. The same principle applies here as in the case of all compounded food products. If they are made in cleanly fashion, contain no harmful ingredients and are truthfully labeled there is no reason for objecting to them. But if the manufacturer goes so far as to put strawberry seeds—or hayseed—into his artificial "strawberry jam" I think that might properly be called adulteration, for it is imitating the imperfections of nature, and man ought to be too proud to do that.

The old-fashioned open kettle molasses consisted mostly of glucose and other invert sugars together with such cane sugar as could not be crystallized out. But when the vacuum pan was introduced the molasses was impoverished of its sweetness and beet sugar does not yield any molasses. So we now have in its place the corn syrups consisting of about 85 per cent. of glucose and 15 per cent. of sugar flavored with maple or vanillin or whatever we like. It is encouraging to see the bill boards proclaiming the virtues of "Karo" syrup and "Mazola" oil when only a few years ago the products of our national cereal were without honor in their own country.

Many other products besides foods are made from corn starch. Dextrin serves in place of the old "gum arabic" for the mucilage of our envelopes and stamps. Another form of dextrin sold as "Kordex" is used to

WHAT COMES FROM CORN

hold together the sand of the cores of castings. After the casting has been made the scorched core can be shaken out. Glucose is used in place of sugar as a filler for cheap soaps and for leather.

Altogether more than a hundred different commercial products are now made from corn, not counting cob pipes. Every year the factories of the United States work up over 50,000,000 bushels of corn into 800,000,000 pounds of corn syrup, 600,000,000 pounds of starch, 230,000,000 pounds of corn sugar, 625,000,000 pounds of gluten feed, 90,000,000 pounds of oil and 90,000,000 pounds of oil cake.

Two million bushels of cobs are wasted every year in the United States. Can't something be made out of them? This is the question that is agitating the chemists of the Carbohydrate Laboratory of the Department of Agriculture at Washington. They have found it possible to work up the corn cobs into glucose and xylose by heating with acid. But glucose can be more cheaply obtained from other starchy or woody materials and they cannot find a market for the xylose. This is a sort of a sugar but only about half as sweet as that from cane. Who can invent a use for it? More promising is the discovery by this laboratory that by digesting the cobs with hot water there can be extracted about 30 per cent. of a gum suitable for bill posting and labeling.

Since the starches and sugars belong to the same class of compounds as the celluloses they also can be acted upon by nitric acid with the production of explosives like guncotton. Nitro-sugar has not come into common use, but nitro-starch is found to be one of safest of the high explosives. On account of the danger of decomposition and spontaneous explosion from the presence of foreign substances the materials in explosives must be of the purest possible. It was formerly thought that tapioca must be imported from Java for making nitro-starch. But during the war when shipping was short, the War Department found that it could be made better and cheaper from our home-grown corn starch. When the war closed the United States was making 1,720,000 pounds of nitro-starch a month for loading hand grenades. So, too, the Post Office Department discovered that it could use mucilage made of corn dextrin as well as that which used to be made from tapioca. This is progress in the right direction. It would be well to divert some of the energetic efforts now devoted to the increase of commerce to the discovery of ways of reducing the need for commerce by the development of home products. There is no merit in simply hauling things around the world.

In the last chapter we saw how dextrose or glucose could be converted by fermentation into alcohol. Since corn starch, as we have here seen, can be converted into dextrose, it can serve as a source of alcohol. This was, in fact, one of the earliest misuses to which corn was put, and before the war put a stop to it 34,000,000 bushels went to the making of whisky in the United States every year, not counting the moonshiners' output. But even though we left off drinking whisky the distillers could still thrive. Mars is more thirsty than Bacchus. The output of alcohol, denatured for industrial purposes, is more than three times what it was before the war, and the price has risen from 30 cents a gallon to 67 cents. This may make it profitable to utilize sugars, starches and cellulose that formerly were out of the question. According to the calculations of the Forest Products Laboratory of Madison it costs from 37 to 44 cents a gallon to make alcohol from corn, but it may be made from sawdust at a cost of from 14 to 20 cents. This is not "wood alcohol" (that is, methyl alcohol, CH_4O) such as is made by the destructive distillation of wood, but genuine "grain alcohol" (ethyl alcohol, C_2H_6O), such as is made by the fermentation of glucose or other sugar. The first step in the process is to digest the sawdust or chips with dilute sulfuric acid under heat and pressure. This converts the cellulose (wood fiber) in large part into glucose ("corn sugar") which may be extracted by hot water in a diffusion battery as in extracting the sugar from beet chips. This glucose solution may then be fermented by yeast and the resulting alcohol distilled off. The process is perfectly practicable but has yet to be proved profitable. But the sulfite liquors of the paper mills are being worked up successfully into industrial alcohol.

The rapidly approaching exhaustion of our oil fields which the war has accelerated leads us to look around to see what we can get to take the place of gasoline. One of the most promising of the suggested substitutes is alcohol. The United States is exceptionally rich in mineral oil, but some countries, for instance England, Germany, France and Australia, have little or none. The Australian Advisory Council of Science, called to consider the problem, recommends alcohol for stationary engines and motor cars. Alcohol has the disadvantage of being less volatile than gasoline so it is hard to start up the engine from the cold. But the lower volatility and ignition point of alcohol are an advantage in that it can be put under a pressure of 150 pounds to the square inch. A pound of gasoline contains fifty per cent. more potential energy than a pound of alcohol, but since the alcohol vapor can be put under twice the compression of the gasoline and requires only one-third the amount of air, the thermal efficiency of an alcohol engine may be fifty per cent. higher than that of a gasoline engine. Alcohol also has several other conveniences that can count in its favor. In the case of incomplete combustion the cylinders are less likely to be clogged with carbon and the escaping gases do not have the offensive odor of the gasoline smoke. Alcohol does not ignite so easily as gasoline and the fire is more readily put out, for water thrown upon blazing alcohol dilutes it and puts out the flame while gasoline floats on water and the fire is spread by it. It is possible to increase the inflammability of alcohol by mixing with it some hydrocarbon such as gasoline, benzene or acetylene. In the Taylor-White process the vapor from low-grade alcohol containing 17 per cent. water is passed over calcium carbide. This takes out the water and adds acetylene gas, making a suitable mixture for an internal combustion engine.

Alcohol can be made from anything of a starchy, sugary or woody nature, that is, from the main substance of all vegetation. If we start with wood (cellulose) we convert it first into sugar (glucose) and, of course, we could stop here and use it for food instead

WHAT COMES FROM CORN

of carrying it on into alcohol. This provides one factor of our food, the carbohydrate, but by growing the yeast plants on glucose and feeding them with nitrates made from the air we can get the protein and fat. So it is quite possible to live on sawdust, although it would be too expensive a diet for anybody but a millionaire, and he would not enjoy it. Glucose has been made from formaldehyde and this in turn made from carbon, hydrogen and oxygen, so the synthetic production of food from the elements is not such an absurdity as it was thought when Berthelot suggested it half a century ago.

The first step in the making of alcohol is to change the starch over into sugar. This transformation is effected in the natural course of sprouting by which the insoluble starch stored up in the seed is converted into the soluble glucose for the sap of the growing plant. This malting process is that mainly made use of in the production of alcohol from grain. But there are other ways of effecting the change. It may be done by heating with acid as we have seen, or according to a method now being developed the final conversion may be accomplished by mold instead of malt. In applying this method, known as the amylo process, to corn, the meal is mixed with twice its weight of water, acidified with hydrochloric acid and steamed. The mash is then cooled down somewhat, diluted with sterilized water and innoculated with the mucor filaments. As the mash molds the starch is gradually changed over to glucose and if this is the product desired the process may be stopped at this point. But if alcohol is wanted yeast is added to ferment the sugar. By keeping it alkaline and treating with the proper bacteria a high yield of glycerin can be obtained.

In the fermentation process for making alcoholic liquors a little glycerin is produced as a by-product. Glycerin, otherwise called glycerol, is intermediate between sugar and alcohol. Its molecule contains three carbon atoms, while glucose has six and alcohol two. It is possible to increase the yield of glycerin if desired by varying the form of fermentation. This was desired most earnestly in Germany during the war, for the British blockade shut off the importation of the fats and oils from which the Germans extracted the glycerin for their nitroglycerin. Under pressure of this necessity they worked out a process of getting glycerin in quantity from sugar and, news of this being brought to this country by Dr. Alonzo Taylor, the United States Treasury Department set up a special laboratory to work out this problem. John R. Eoff and other chemists working in this laboratory succeeded in getting a yield of twenty per cent. of glycerin by fermenting black strap molasses or other syrup with California wine yeast. During the fermentation it is necessary to neutralize the acetic acid formed with sodium or calcium carbonate. It was estimated that glycerin could be made from waste sugars at about a quarter of its war-time cost, but it is doubtful whether the process would be profitable at normal prices.

We can, if we like, dispense with either yeast or bacteria in the production of glycerin. Glucose syrup suspended in oil under steam pressure with finely divided nickel as a catalyst and treated with nascent hydrogen

WHAT COMES FROM CORN

will take up the hydrogen and be converted into glycerin. But the yield is poor and the process expensive.

Food serves substantially the same purpose in the body as fuel in the engine. It provides the energy for work. The carbohydrates, that is the sugars, starches and celluloses, can all be used as fuels and can all— even, as we have seen, the cellulose—be used as foods. The final products, water and carbon dioxide, are in both cases the same and necessarily therefore the amount of energy produced is the same in the body as in the engine. Corn is a good example of the equivalence of the two sources of energy. There are few better foods and no better fuels. I can remember the good old days in Kansas when we had corn to burn. It was both an economy and a luxury, for—at ten cents a bushel—it was cheaper than coal or wood and preferable to either at any price. The long yellow ears, each wrapped in its own kindling, could be handled without crocking the fingers. Each kernel as it crackled sent out a blazing jet of oil and the cobs left a fine bed of coals for the corn popper to be shaken over. Driftwood and the pyrotechnic fuel they make now by soaking sticks in strontium and copper salts cannot compare with the old-fashioned corn-fed fire in beauty and the power of evoking visions. Doubtless such luxury would be condemned as wicked nowadays, but those who have known the calorific value of corn would find it hard to abandon it altogether, and I fancy that the Western farmer's wife, when she has an extra batch of baking to do, will still steal a few ears from the crib.

XI

SOLIDIFIED SUNSHINE

All life and all that life accomplishes depend upon the supply of solar energy stored in the form of food. The chief sources of this vital energy are the fats and the sugars. The former contain two and a quarter times the potential energy of the latter. Both, when completely purified, consist of nothing but carbon, hydrogen and oxygen; elements that are to be found freely everywhere in air and water. So when the sunny southland exports fats and oils, starches and sugar, it is then sending away nothing material but what comes back to it in the next wind. What it is sending to the regions of more slanting sunshine is merely some of the surplus of the radiant energy it has received so abundantly, compacted for convenience into a portable and edible form.

In previous chapters I have dealt with some of the uses of cotton, its employment for cloth, for paper, for artificial fibers, for explosives, and for plastics. But I have ignored the thing that cotton is attached to and for which, in the economy of nature, the fibers are formed; that is, the seed. It is as though I had described the aeroplane and ignored the aviator whom it was designed to carry. But in this neglect I am but following the example of the human race, which for three thousand years used the fiber but made no use of the seed except to plant the next crop.

SOLIDIFIED SUNSHINE

Just as mankind is now divided into the two great classes, the wheat-eaters and the rice-eaters, so the ancient world was divided into the wool-wearers and the cotton-wearers. The people of India wore cotton; the Europeans wore wool. When the Greeks under Alexander fought their way to the Far East they were surprised to find wool growing on trees. Later travelers returning from Cathay told of the same marvel and travelers who stayed at home and wrote about what they had not seen, like Sir John Maundeville, misunderstood these reports and elaborated a legend of a tree that bore live lambs as fruit. Here, for instance, is how a French poetical botanist, Delacroix, described it in 1791, as translated from his Latin verse:

> Upon a stalk is fixed a living brute,
> A rooted plant bears quadruped for fruit;
> It has a fleece, nor does it want for eyes,
> And from its brows two wooly horns arise.
> The rude and simple country people say
> It is an animal that sleeps by day
> And wakes at night, though rooted to the ground,
> To feed on grass within its reach around.

But modern commerce broke down the barrier between East and West. A new cotton country, the best in the world, was discovered in America. Cotton invaded England and after a hard fight, with fists as well as finance, wool was beaten in its chief stronghold. Cotton became King and the wool-sack in the House of Lords lost its symbolic significance.

Still two-thirds of the cotton crop, the seed, was wasted and it is only within the last fifty years that methods of using it have been developed to any extent.

The cotton crop of the United States for 1917 amounted to about 11,000,000 bales of 500 pounds each. When the Great War broke out and no cotton could be exported to Germany and little to England the South was in despair, for cotton went down to five or six cents a pound. The national Government, regardless of states' rights, was called upon for aid and everybody was besought to "buy a bale." Those who responded to this patriotic appeal were well rewarded, for cotton

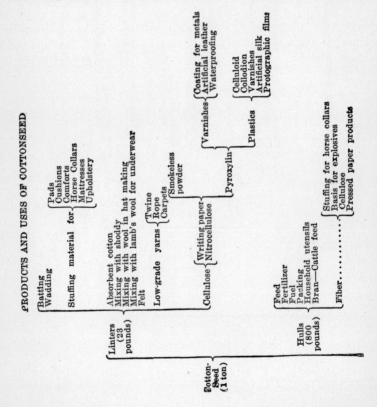

rose as the war went on and sold at twenty-nine cents a pound.

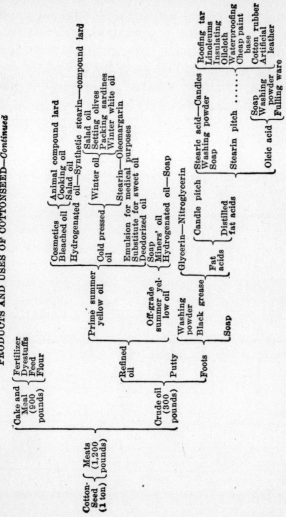

But the chemist has added some $150,000,000 a year to the value of the crop by discovering ways of utilizing the cottonseed that used to be thrown away or burned as fuel. The genealogical table of the progeny of the cottonseed herewith printed will give some idea of their variety. If you will examine a cottonseed you will see first that there is a fine fuzz of cotton fiber sticking to it. These linters can be removed by machinery and used for any purpose where length of fiber is not essential. For instance, they may be nitrated as described in previous articles and used for making smokeless powder or celluloid.

On cutting open the seed you will observe that it consists of an oily, mealy kernel encased in a thin brown hull. The hulls, amounting to 700 or 900 pounds in a ton of seed, were formerly burned. Now, however, they bring from $4 to $10 a ton because they can be ground up into cattle-feed or paper stock or used as fertilizer.

The kernel of the cottonseed on being pressed yields a yellow oil and leaves a mealy cake. This last, mixed with the hulls, makes a good fodder for fattening cattle. Also, adding twenty-five per cent. of the refined cottonseed meal to our war bread made it more nutritious and no less palatable. Cottonseed meal contains about forty per cent. of protein and is therefore a highly concentrated and very valuable feeding stuff. Before the war we were exporting nearly half a million tons of cottonseed meal to Europe, chiefly to Germany and Denmark, where it is used for dairy cows. The British yeoman, his country's pride, has not yet been won over to the use of any such newfangled fodder and consequently the British manufacturer could not compete with his continental rivals in the seed-crushing business, for he could not dispose of his meal-cake by-product as did they.

Let us now turn to the most valuable of the cottonseed products, the oil. The seed contains about twenty per cent. of oil, most of which can be squeezed out of the hot seeds by hydraulic pressure. It comes out as a red liquid of a disagreeable odor. This is decolorized, deodorized and otherwise purified in various ways: by treatment with alkalies or acids, by blowing air and steam through it, by shaking up with fuller's earth, by settling and filtering. The refined product is a yellow oil, suitable for table use. Formerly, on account of the popular prejudice against any novel food products, it used to masquerade as olive oil. Now, however, it boldly competes with its ancient rival in the lands of the olive tree and America ships some 700,000 barrels of cottonseed oil a year to the Mediterranean. The Turkish Government tried to check the spread of cottonseed oil by calling it an adulterant and prohibiting its mixture with olive oil. The result was that the sale of Turkish olive oil fell off because people found its flavor too strong when undiluted. Italy imports cottonseed oil and exports her olive oil. Denmark imports cottonseed meal and margarine and exports her butter.

Northern nations are accustomed to hard fats and do not take to oils for cooking or table use as do the southerners. Butter and lard are preferred to olive oil and ghee. But this does not rule out cottonseed. It can be combined with the hard fats of animal or vegetable origin in margarine or it may itself be hardened by hydrogen.

To understand this interesting reaction which is profoundly affecting international relations it will be necessary to dip into the chemistry of the subject. Here are the symbols of the chief ingredients of the fats and oils. Please look at them.

$$\text{Linoleic acid} \ldots\ldots\ldots\ldots\ldots\ldots\ldots\ldots C_{18}H_{32}O_2$$
$$\text{Oleic acid} \ldots\ldots\ldots\ldots\ldots\ldots\ldots\ldots\ldots C_{18}H_{34}O_2$$
$$\text{Stearic acid} \ldots\ldots\ldots\ldots\ldots\ldots\ldots\ldots C_{18}H_{36}O_2$$

Don't skip these because you have not studied chemistry. That's why I am giving them to you. If you had studied chemistry you would know them without my telling. Just examine them and you will discover the secret. You will see that all three are composed of the same elements, carbon, hydrogen and oxygen. Notice next the number of atoms of each element as indicated by the little low figures on the right of each letter. You observe that all three contain the same number of atoms of carbon and oxygen but differ in the amount of hydrogen. This trifling difference in composition makes a great difference in behavior. The less the hydrogen the lower the melting point. Or to say the same thing in other words, fatty substances low in hydrogen are apt to be liquids and those with a full complement of hydrogen atoms are apt to be solids at the ordinary temperature of the air. It is common to call the former "oils" and the latter "fats," but that implies too great a dissimilarity, for the distinction depends on whether we are living in the tropics or the arctic. It is better, therefore, to lump them all together and call them "soft fats" and "hard fats," respectively.

Fats of the third order, the stearic group, are called "saturated" because they have taken up all the hydrogen they can hold. Fats of the other two groups are called "unsaturated." The first, which have the least hydrogen, are the most eager for more. If hydrogen is not handy they will take up other things, for instance oxygen. Linseed oil, which consists largely, as the name implies, of linoleic acid, will absorb oxygen on exposure to the air and become hard. That is why it is used in painting. Such oils are called "drying" oils, although the hardening process is not really drying, since they contain no water, but is oxidation. The "semi-drying oils," those that will harden somewhat on exposure to the air, include the oils of cottonseed, corn, sesame, soy bean and castor bean. Olive oil and peanut oil are "non-drying" and contain oleic compounds (olein). The hard fats, such as stearin, palmitin and margarin, are mostly of animal origin, tallow and lard, though coconut and palm oil contain a large proportion of such saturated compounds.

Though the chemist talks of the fatty "acids," nobody else would call them so because they are not sour. But they do behave like the acids in forming salts with bases. The alkali salts of the fatty acids are known to us as soaps. In the natural fats they exist not as free acids but as salts of an organic base, glycerin, as I explained in a previous chapter. The natural fats and oils consist of complex mixtures of the glycerin compounds of these acids (known as olein, stearin, etc.), as well as various others of a similar sort. If you will set a bottle of salad oil in the ice-box you will see it separate into two parts. The white, crystalline solid that separates out is largely stearin. The part that remains liquid is largely olein. You might separate them by filtering it cold and if then you tried to sell the two products you would find that the hard fat would bring a higher price than the oil, either for food or soap. If you tried to keep them you would find that the hard fat kept neutral and "sweet" longer than the other. You may remember that the perfumes (as well as their odorous opposites) were mostly unsaturated compounds. So we find that it is the free and unsaturated fatty acids that cause butter and oil to become rank and rancid.

Obviously, then, we could make money if we could turn soft, unsaturated fats like olein into hard, saturated fats like stearin. Referring to the symbols we see that all that is needed to effect the change is to get the former to unite with hydrogen. This requires a little coaxing. The coaxer is called a catalyst. A catalyst, as I have previously explained, is a substance that by its mere presence causes the union of two other substances that might otherwise remain separate. For that reason the catalyst is referred to as "a chemical parson." Finely divided metals have a strong catalytic action. Platinum sponge is excellent but too expensive. So in this case nickel is used. A nickel salt mixed with charcoal or pumice is reduced to the metallic state by heating in a current of hydrogen. Then it is dropped into the tank of oil and hydrogen gas is blown through. The hydrogen may be obtained by splitting water into its two components, hydrogen and oxygen, by means of the electrical current, or by passing steam over spongy iron which takes out the oxygen. The stream of hydrogen blown through the hot oil converts the linoleic acid to oleic and then the oleic into stearic. If you figured up the weights from the symbols given above you would find that it takes about one pound of hydrogen to convert a hundred pounds of olein to stearin and the cost is only about one cent a pound. The nickel is unchanged and is easily separated. A trace of nickel may remain in the product, but as it is very much less than the amount dissolved when food is cooked in nickel-plated vessels it cannot be regarded as harmful.

Even more unsaturated fats may be hydrogenated. Fish oil has hitherto been almost unusable because of its powerful and persistent odor. This is chiefly due to a fatty acid which properly bears the uneuphonious name of clupanodonic acid and has the composition of $C_{18}H_{28}O_2$. By comparing this with the symbol of the odorless stearic acid, $C_{18}H_{36}O_2$, you will see that all the rank fish oil lacks to make it respectable is eight hydrogen atoms. A Japanese chemist, Tsujimoto, has discovered how to add them and now the reformed fish oil under the names of "talgol" and "candelite" serves for lubricant and even enters higher circles as a soap or food.

This process of hardening fats by hydrogenation resulted from the experiments of a French chemist, Professor Sabatier of Toulouse, in the last years of the last century, but, as in many other cases, the Germans were the first to take it up and profit by it. Before the war the copra or coconut oil from the British Asiatic colonies of India, Ceylon and Malaya went to Germany at the rate of $15,000,000 a year. The palm kernels grown in British West Africa were shipped, not to Liverpool, but to Hamburg, $19,000,000 worth annually. Here the oil was pressed out and used for margarin and the residual cake used for feeding cows produced butter or for feeding hogs produced lard. Half of the copra raised in the British possessions was sent to Germany and half of the oil from it was resold to the British margarin candle and soap makers at a handsome profit. The British chemists were not blind to this, but they could do nothing, first because the English politician was wedded to free trade, second, because the English farmer would not use oil cake for his stock. France was in a similar situation. Marseilles produced 15,-500,000 gallons of oil from peanuts grown largely in the French African colonies—but shipped the oil-cake on to Hamburg. Meanwhile the Germans, in pursuit of their policy of attaining economic independence, were striving to develop their own tropical territory. The subjects of King George who because they had the misfortune to live in India were excluded from the British South African dominions or mistreated when they did come, were invited to come to German East Africa and set to raising peanuts in rivalry to French Senegal and British Coromandel. Before the war Germany got half of the Egyptian cottonseed and half of the Philippine copra. That is one of the reasons why German warships tried to check Dewey at Manila in 1898 and German troops tried to conquer Egypt in 1915.

But the tide of war set the other way and the German plantations of palmnuts and peanuts in Africa have come into British possession and now the British Government is starting an educational campaign to teach their farmers to feed oil cake like the Germans and their people to eat peanuts like the Americans.

The Germans shut off from the tropical fats supply were hard up for food and for soap, for lubricants and for munitions. Every person was given a fat card that reduced his weekly allowance to the minimum. Millers were required to remove the germs from their cereals and deliver them to the war department. Children were set to gathering horse-chestnuts, elderberries, linden-balls, grape seeds, cherry stones and sunflower heads, for these contain from six to twenty per cent. of oil. Even the blue-bottle fly—hitherto an idle creature for whom Beelzebub found mischief—was conscripted into the national service and set to laying eggs by the billion on fish refuse. Within a few days there is a crop of larvæ which, to quote the "Chemische Zentralblatt," yields forty-five grams per kilogram of a yellow oil. This product, we should hope, is used for axle-grease and nitroglycerin, although properly purified it would be as nutritious as any other—to one who has no imagination. Driven to such straits Germany would have given a good deal for one of those tropical islands that we are so careless about.

It might have been supposed that since the United States possessed the best land in the world for the production of cottonseed, coconuts, peanuts and corn that it would have led all other countries in the utilization of vegetable oils for food. That this country has not so used its advantage is due to the fact that the new products have not merely had to overcome popular conservatism, ignorance and prejudice—hard things to fight in any case—but have been deliberately checked and hampered by the state and national governments in defense of vested interests. The farmer vote is a power that no politician likes to defy and the dairy business in every state was thoroughly organized. In New York the oleomargarin industry that in 1879 was turning out products valued at more than $5,000,000 a year was completely crushed out by state legislation.[1] The output of the United States, which in 1902 had risen to 126,000,000 pounds, was cut down to 43,000,000 pounds in 1909 by federal legislation. According to the disingenuous custom of American lawmakers the Act of 1902 was passed through Congress as a "revenue measure," although it meant a loss to the Government of more than three million dollars a year over what might be produced by a straight two cents a pound tax. A wholesale dealer in oleomargarin was made to pay a higher license than a wholesale liquor dealer. The federal law put a tax of ten cents a pound on yellow oleomargarin and a quarter of a cent a pound on the uncolored. But people—doubtless from pure prejudice—prefer a yellow spread for their bread, so the economical housewife has to work over her oleomargarin with the annatto which is given to her when she buys a package or, if the law prohibits this, which she is permitted to steal from an open box on the grocer's counter. A plausible pretext for such legislation is afforded by the fact that the butter substitutes are so much like butter that they cannot be easily distinguished from it unless the use of annatto is permitted

[1] United States Abstract of Census of Manufactures, 1914, p. 34.

SOLIDIFIED SUNSHINE

to butter and prohibited to its competitors. Fradulent sales of substitutes of any kind ought to be prevented, but the recent pure food legislation in America has shown that it is possible to secure truthful labeling without resorting to such drastic measures. In Europe the laws against substitution were very strict, but not devised to restrict the industry. Consequently the margarin output of Germany doubled in the five years preceding the war and the output of England tripled. In Denmark the consumption of margarin rose from 8.8 pounds per capita in 1890 to 32.6 pounds in 1912. Yet the butter business, Denmark's pride, was not injured, and Germany and England imported more butter than ever before. Now that the price of butter in America has gone over the seventy-five cent mark Congress may conclude that it no longer needs to be protected against competition.

The "compound lards" or "lard compounds," consisting usually of cottonseed oil and oleo-stearin, although the latter may now be replaced by hardened oil, met with the same popular prejudice and attempted legislative interference, but succeeded more easily in coming into common use under such names as "Cottosuet," "Kream Krisp," "Kuxit," "Korno," "Cottolene" and "Crisco."

Oleomargarin, now generally abbreviated to margarin, originated, like many other inventions, in military necessity. The French Government in 1869 offered a prize for a butter substitute for the army that should be cheaper and better than butter in that it did not spoil so easily. The prize was won by a French chemist, Mége-Mouries, who found that by chilling beef fat the solid stearin could be separated from an oil (oleo) which was the substantially same as that in milk and hence in butter. Neutral lard acts the same.

This discovery of how to separate the hard and soft fats was followed by improved methods for purifying them and later by the process for converting the soft into the hard fats by hydrogenation. The net result was to put into the hands of the chemist the ability to draw his materials at will from any land and from the vegetable and animal kingdoms and to combine them as he will to make new fat foods for every use; hard for summer, soft for winter; solid for the northerners and liquid for the southerners; white, yellow or any other color, and flavored to suit the taste. The Hindu can eat no fat from the sacred cow; the Mohammedan and the Jew can eat no fat from the abhorred pig; the vegetarian will touch neither; other people will take both. No matter, all can be accommodated.

All the fats and oils, though they consist of scores of different compounds, have practically the same food value when freed from the extraneous matter that gives them their characteristic flavors. They are all practically tasteless and colorless. The various vegetable and animal oils and fats have about the same digestibility, 98 per cent.,[1] and are all ordinarily completely utilized in the body, supplying it with two and a quarter times as much energy as any other food.

It does not follow, however, that there is no difference in the products. The margarin men accuse butter of harboring tuberculosis germs from which their product, because it has been heated or is made from vege-

[1] United States Department of Agriculture, Bulletin No. 505.

table fats, is free. The butter men retort that margarin is lacking in vitamines, those mysterious substances which in minute amounts are necessary for life and especially for growth. Both the claim and the objection lose a large part of their force where the margarin, as is customarily the case, is mixed with butter or churned up with milk to give it the familiar flavor. But the difficulty can be easily overcome. The milk used for either butter or margarin should be free or freed from disease germs. If margarin is altogether substituted for butter, the necessary vitamines may be sufficiently provided by milk, eggs and greens.

Owing to these new processes all the fatty substances of all lands have been brought into competition with each other. In such a contest the vegetable is likely to beat the animal and the southern to win over the northern zones. In Europe before the war the proportion of the various ingredients used to make butter substitutes was as follows:

AVERAGE COMPOSITION OF EUROPEAN MARGARIN

		Per Cent.
Animal hard fats		25
Vegetable hard fats		35
Copra	29	
Palm-kernel	6	
Vegetable soft fats		26
Cottonseed	13	
Peanut	6	
Sesame	6	
Soya-bean	1	
Water, milk, salt		14
		100

This is not the composition of any particular brand but the average of them all. The use of a certain amount of the oil of the sesame seed is required by the laws of Germany and Denmark because it can be easily detected by a chemical color test and so serves to prevent the margarin containing it from being sold as butter. "Open sesame!" is the password to these markets. Remembering that margarin originally was made up entirely of animal fats, soft and hard, we can see from the above figures how rapidly they are being displaced by the vegetable fats. The cottonseed and peanut oils have replaced the original oleo oil and the tropical oils from the coconut (copra) and African palm are crowding out the animal hard fats. Since now we can harden at will any of the vegetable oils it is possible to get along altogether without animal fats. Such vegetable margarins were originally prepared for sale in India, but proved unexpectedly popular in Europe, and are now being introduced into America. They are sold under various trade names suggesting their origin, such as "palmira," "palmona," "milkonut," "cocose," "coconut oleomargarin" and "nucoa nut margarin." The last named is stated to be made of coconut oil (for the hard fat) and peanut oil (for the soft fat), churned up with a culture of pasteurized milk (to impart the butter flavor). The law requires such a product to be branded "oleomargarine" although it is not. Such cases of compulsory mislabeling are not rare. You remember the "Pigs is Pigs" story.

Peanut butter has won its way into the American menu without any camouflage whatever, and as a salad oil it is almost equally frank about its lowly origin.

This nut, which grows on a vine instead of a tree, and is dug from the ground like potatoes instead of being picked with a pole, goes by various names according to locality, peanuts, ground-nuts, monkey-nuts, arachides and goobers. As it takes the place of cotton oil in some of its products so it takes its place in the fields and oilmills of Texas left vacant by the bollweevil. The once despised peanut added some $56,000,000 to the wealth of the South in 1916. The peanut is rich in the richest of foods, some 50 per cent. of oil and 30 per cent. of protein. The latter can be worked up into meat substitutes that will make the vegetarian cease to envy his omnivorous neighbor. Thanks largely to the chemist who has opened these new fields of usefulness, the peanut-raiser got $1.25 a bushel in 1917 instead of the 30 cents that he got four years before.

It would be impossible to enumerate all the available sources of vegetable oils, for all seeds and nuts contain more or less fatty matter and as we become more economical we shall utilize of what we now throw away. The germ of the corn kernel, once discarded in the manufacture of starch, now yields a popular table oil. From tomato seeds, one of the waste products of the canning factory, can be extracted 22 per cent. of an edible oil. Oats contain 7 per cent. of oil. From rape seed the Japanese get 20,000 tons of oil a year. To the sources previously mentioned may be added pumpkin seeds, poppy seeds, raspberry seeds, tobacco seeds, cockleburs, hazelnuts, walnuts, beechnuts and acorns.

The oil-bearing seeds of the tropics are innumerable and will become increasingly essential to the inhabitants of northern lands. It was the realization of this that brought on the struggle of the great powers for the possession of tropical territory which, for years before, they did not think worth while raising a flag over. No country in the future can consider itself safe unless it has secure access to such sources. We had a sharp lesson in this during the war. Palm oil, it seems, is necessary for the manufacture of tinplate, an industry that was built up in the United States by the McKinley tariff. The British possessions in West Africa were the chief source of palm oil and the Germans had the handling of it. During the war the British Government assumed control of the palm oil products of the British and German colonies and prohibited their export to other countries than England. Americans protested and beseeched, but in vain. The British held, quite correctly, that they needed all the oil they could get for food and lubrication and nitroglycerin. But the British also needed canned meat from America for their soldiers and when it was at length brought to their attention that the packers could not ship meat unless they had cans and that cans could not be made without tin and that tin could not be made without palm oil the British Government consented to let us buy a little of their palm oil. The lesson is that of Voltaire's story, "Candide," "Let us cultivate our own garden"—and plant a few palm trees in it—also rubber trees, but that is another story.

The international struggle for oil led to the partition of the Pacific as the struggle for rubber led to the partition of Africa. Theodor Weber, as Stevenson says, "harried the Samoans" to get copra much as King Leopold of Belgium harried the Congoese to get

SOLIDIFIED SUNSHINE

caoutchouc. It was Weber who first fully realized that the South Sea islands, formerly given over to cannibals, pirates and missionaries, might be made immensely valuable through the cultivation of the coconut palms. When the ripe coconut is split open and exposed to the sun the meat dries up and shrivels and in this form, called "copra," it can be cut out and shipped to the factory where the oil is extracted and refined. Weber while German Consul in Samoa was also manager of what was locally known as "the long-handled concern" (*Deutsche Handels und Plantagen Gesellschaft der Südsee Inseln zu Hamburg*), a pioneer commercial and semi-official corporation that played a part in the Pacific somewhat like the British Hudson Bay Company in Canada or East India Company in Hindustan. Through the agency of this corporation on the start Germany acquired a virtual monopoly of the transportation and refining of coconut oil and would have become the dominant power in the Pacific if she had not been checked by force of arms. In Apia Bay in 1889 and again in Manila Bay in 1898 an American fleet faced a German fleet ready for action while a British warship lay between. So we rescued the Philippines and Samoa from German rule and in 1914 German power was eliminated from the Pacific. During the ten years before the war, the production of copra in the German islands more than doubled and this was only the beginning of the business. Now these islands have been divided up among Australia, New Zealand and Japan, and these countries are planning to take care of the copra.

But although we get no extension of territory from

the war we still have the Philippines and some of the Samoan Islands, and these are capable of great development. From her share of the Samoan Islands Germany got a million dollars' worth of copra and we might get more from ours. The Philippines now lead the world in the production of copra, but Java is a close second and Ceylon not far behind. If we do not look out we will be beaten both by the Dutch and the British, for they are undertaking the cultivation of the coconut on a larger scale and in a more systematic way. According to an official bulletin of the Philippine Government a coconut plantation should bring in "dividends ranging from 10 to 75 per cent. from the tenth to the hundredth year." And this being printed in 1913 figured the price of copra at 3½ cents, whereas it brought 4½ cents in 1918, so the prospect is still more encouraging. The copra is half fat and can be cheaply shipped to America, where it can be crushed in the southern oil-mills when they are not busy on cottonseed or peanuts. But even this cost of transportation can be reduced by extracting the oil in the islands and shipping it in bulk like petroleum in tank steamers.

In the year ending June, 1918, the United States imported from the Philippines 155,000,000 pounds of coconut oil worth $18,000,000 and 220,000,000 pounds of copra worth $10,000,000. But this was about half our total importations; the rest of it we had to get from foreign countries. Panama palms may give us a little relief from this dependence on foreign sources. In 1917 we imported 19,000,000 whole coconuts from Panama valued at $700,000.

A new form of fat that has rapidly come into our

market is the oil of the soya or soy bean. In 1918 we imported over 300,000,000 pounds of soy-bean oil, mostly from Manchuria. The oil is used in manufacture of substitutes for butter, lard, cheese, milk and cream, as well as for soap and paint. The soy-bean can be raised in the United States wherever corn can be grown and provides provender for man and beast. The soy meal left after the extraction of the oil makes a good cattle food and the fermented juice affords the shoya sauce made familiar to us through the popularity of the chop-suey restaurants.

As meat and dairy products become scarcer and dearer we shall become increasingly dependent upon the vegetable fats. We should therefore devise means of saving what we now throw away, raise as much as we can under our own flag, keep open avenues for our foreign supply and encourage our cooks to make use of the new products invented by our chemists.

CHAPTER XII

FIGHTING WITH FUMES

The Germans opened the war using projectiles seventeen inches in diameter. They closed it using projectiles one one-hundred millionth of an inch in diameter. And the latter were more effective than the former. As the dimensions were reduced from molar to molecular the battle became more intense. For when the Big Bertha had shot its bolt, that was the end of it. Whomever it hit was hurt, but after that the steel fragments of the shell lay on the ground harmless and inert. The men in the dugouts could hear the shells whistle overhead without alarm. But the poison gas could penetrate where the rifle ball could not. The malignant molecules seemed to search out their victims. They crept through the crevices of the subterranean shelters. They hunted for the pinholes in the face masks. They lay in wait for days in the trenches for the soldiers' return as a cat watches at the hole of a mouse. The cannon ball could be seen and heard. The poison gas was invisible and inaudible, and sometimes even the chemical sense which nature has given man for his protection, the sense of smell, failed to give warning of the approach of the foe.

The smaller the matter that man can deal with the more he can get out of it. So long as man was dependent for power upon wind and water his working capac-

FIGHTING WITH FUMES

ity was very limited. But as soon as he passed over the border line from physics into chemistry and learned how to use the molecule, his efficiency in work and warfare was multiplied manifold. The molecular bombardment of the piston by steam or the gases of combustion runs his engines and propels his cars. The first man who wanted to kill another from a safe distance threw the stone by his arm's strength. David added to his arm the centrifugal force of a sling when he slew Goliath. The Romans improved on this by concentrating in a catapult the strength of a score of slaves and casting stone cannon balls to the top of the city wall. But finally man got closer to nature's secret and discovered that by loosing a swarm of gaseous molecules he could throw his projectile seventy-five miles and then by the same force burst it into flying fragments. There is no smaller projectile than the atom unless our belligerent chemists can find a way of using the electron stream of the cathode ray. But this so far has figured only in the pages of our scientific romancers and has not yet appeared on the battlefield. If, however, man could tap the reservoir of sub-atomic energy he need do no more work and would make no more war, for unlimited powers of construction and destruction would be at his command. The forces of the infinitesimal are infinite.

The reason why a gas is so active is because it is so egoistic. Psychologically interpreted, a gas consists of particles having the utmost aversion to one another. Each tries to get as far away from every other as it can. There is no cohesive force; no attractive impulse; nothing to draw them together except the all too feeble power of gravitation. The hotter they get the more they try to disperse and so the gas expands. The gas represents the extreme of individualism as steel represents the extreme of collectivism. The combination of the two works wonders. A hot gas in a steel cylinder is the most powerful agency known to man, and by means of it he accomplishes his greatest achievements in peace or war time.

The projectile is thrown from the gun by the expansive force of the gases released from the powder and when it reaches its destination it is blown to pieces by the same force. This is the end of it if it is a shell of the old-fashioned sort, for the gases of combustion mingle harmlessly with the air of which they are normal constituents. But if it is a poison gas shell each molecule as it is released goes off straight into the air with a speed twice that of the cannon ball and carries death with it. A man may be hit by a heavy piece of lead or iron and still survive, but an unweighable amount of lethal gas may be fatal to him.

Most of the novelties of the war were merely extensions of what was already known. To increase the caliber of a cannon from 38 to 42 centimeters or its range from 30 to 75 miles does indeed make necessary a decided change in tactics, but it is not comparable to the revolution effected by the introduction of new weapons of unprecedented power such as airplanes, submarines, tanks, high explosives or poison gas. If any army had been as well equipped with these in the beginning as all armies were at the end it might easily have won the war. That is to say, if the general staff of any of the powers had had the foresight and confidence to develop

FIGHTING WITH FUMES

and practise these modes of warfare on a large scale in advance it would have been irresistible against an enemy unprepared to meet them. But no military genius appeared on either side with sufficient courage and imagination to work out such schemes in secret before trying them out on a small scale in the open. Consequently the enemy had fair warning and ample time to learn how to meet them and methods of defense developed concurrently with methods of attack. For instance, consider the motor fortresses to which Ludendorff ascribes his defeat. The British first sent out a few clumsy tanks against the German lines. Then they set about making a lot of stronger and livelier ones, but by the time these were ready the Germans had field guns to smash them and chain fences with concrete posts to stop them. On the other hand, if the Germans had followed up their advantage when they first set the cloud of chlorine floating over the battlefield of Ypres they might have won the war in the spring of 1915 instead of losing it in the fall of 1918. For the British were unprepared and unprotected against the silent death that swept down upon them on the 22nd of April, 1915. What happened then is best told by Sir Arthur Conan Doyle in his "History of the Great War."

From the base of the German trenches over a considerable length there appeared jets of whitish vapor, which gathered and swirled until they settled into a definite low cloud-bank, greenish-brown below and yellow above, where it reflected the rays of the sinking sun. This ominous bank of vapor, impelled by a northern breeze, drifted swiftly across the space which separated the two lines. The French troops, staring over the top of their parapet at this curious screen which ensured them a temporary relief from fire, were observed suddenly to throw up their hands, to clutch at their throats, and to fall to the ground in the agonies of asphyxiation. Many lay where they had fallen, while their comrades, absolutely helpless against this diabolical agency, rushed madly out of the mephitic mist and made for the rear, over-running the lines of trenches behind them. Many of them never halted until they had reached Ypres, while others rushed westwards and put the canal between themselves and the enemy. The Germans, meanwhile, advanced, and took possession of the successive lines of trenches, tenanted only by the dead garrisons, whose blackened faces, contorted figures, and lips fringed with the blood and foam from their bursting lungs, showed the agonies in which they had died. Some thousands of stupefied prisoners, eight batteries of French field-guns, and four British 4.7's, which had been placed in a wood behind the French position, were the trophies won by this disgraceful victory.

Under the shattering blow which they had received, a blow particularly demoralizing to African troops, with their fears of magic and the unknown, it was impossible to rally them effectually until the next day. It is to be remembered in explanation of this disorganization that it was the first experience of these poison tactics, and that the troops engaged received the gas in a very much more severe form than our own men on the right of Langemarck. For a time there was a gap five miles broad in the front of the position of the Allies, and there were many hours during which there was no substantial force between the Germans and Ypres. They wasted their time, however, in consolidating their ground, and the chance of a great coup passed forever. They had sold their souls as soldiers, but the Devil's price was a poor one. Had they had a corps of cavalry ready, and pushed them through the gap, it would have been the most dangerous moment of the war.

FIGHTING WITH FUMES

A deserter had come over from the German side a week before and told them that cylinders of poison gas had been laid in the front trenches, but no one believed him or paid any attention to his tale. War was then, in the Englishman's opinion, a gentleman's game, the royal sport, and poison was prohibited by the Hague rules. But the Germans were not playing the game according to the rules, so the British soldiers were strangled in their own trenches and fell easy victims to the advancing foe. Within half an hour after the gas was turned on 80 per cent. of the opposing troops were knocked out. The Canadians, with wet handkerchiefs over their faces, closed in to stop the gap, but if the Germans had been prepared for such success they could have cleared the way to the coast. But after such trials the Germans stopped the use of free chlorine and began the preparation of more poisonous gases. In some way that may not be revealed till the secret history of the war is published, the British Intelligence Department obtained a copy of the lecture notes of the instructions to the German staff giving details of the new system of gas warfare to be started in December. Among the compounds named was phosgene, a gas so lethal that one part in ten thousand of air may be fatal. The antidote for it is hexamethylene tetramine. This is not something the soldier—or anybody else—is accustomed to carry around with him, but the British having had a chance to cram up in advance on the stolen lecture notes were ready with gas helmets soaked in the reagent with the long name.

The Germans rejoiced when gas bombs took the place of bayonets because this was a field in which intelligence counted for more than brute force and in which therefore they expected to be supreme. As usual they were right in their major premise but wrong in their conclusion, owing to the egoism of their implicit minor premise. It does indeed give the advantage to skill and science, but the Germans were beaten at their own game, for by the end of the war the United States was able to turn out toxic gases at a rate of 200 tons a day, while the output of Germany or England was only about 30 tons. A gas plant was started at Edgewood, Maryland, in November, 1917. By March it was filling shell and before the war put a stop to its activities in the fall it was producing 1,300,000 pounds of chlorin, 1,000,000 pounds of chlorpicrin, 1,300,000 pounds of phosgene and 700,000 pounds of mustard gas a month.

Chlorine, the first gas used, is unpleasantly familiar to every one who has entered a chemical laboratory or who has smelled the breath of bleaching powder. It is a greenish-yellow gas made from common salt. The Germans employed it at Ypres by laying cylinders of the liquefied gas in the trenches, about a yard apart, and running a lead discharge pipe over the parapet. When the stop cocks are turned the gas streams out and since it is two and a half times as heavy as air it rolls over the ground like a noisome mist. It works best when the ground slopes gently down toward the enemy and when the wind blows in that direction at a rate between four and twelve miles an hour. But the wind, being strictly neutral, may change its direction without warning and then the gases turn back in their flight and attack their own side, something that rifle bullets have never been known to do.

FIGHTING WITH FUMES

Because free chlorine would not stay put and was dependent on the favor of the wind for its effect, it was later employed, not as an elemental gas, but in some volatile liquid that could be fired in a shell and so released at any particular point far back of the front trenches.

The most commonly used of these compounds was phosgene, which, as the reader can see by inspection of its formula, $COCl_2$, consists of chlorine (Cl) combined with carbon monoxide (CO), the cause of deaths from illuminating gas. These two poisonous gases, chlorine and carbon monoxide, when mixed together, will not readily unite, but if a ray of sunlight falls upon the mixture they combine at once. For this reason John Davy, who discovered the compound over a hundred years ago, named it phosgene, that is, "produced by light." The same roots recur in hydrogen, so named because it is "produced from water," and phosphorus, because it is a "light-bearer."

In its modern manufacture the catalyzer or instigator of the combination is not sunlight but porous carbon. This is packed in iron boxes eight feet long, through which the mixture of the two gases was forced. Carbon monoxide may be made by burning coke with a supply of air insufficient for complete combustion, but in order to get the pure gas necessary for the phosgene common air was not used, but instead pure oxygen extracted from it by a liquid air plant.

Phosgene is a gas that may be condensed easily to a liquid by cooling it down to 46 degrees Fahrenheit. A mixture of three-quarters chlorine with one-quarter phosgene has been found most effective. By itself phosgene has an inoffensive odor somewhat like green corn and so may fail to arouse apprehension until a toxic concentration is reached. But even small doses have such an effect upon the heart action for days afterward that a slight exertion may prove fatal.

The compound manufactured in largest amount in America was chlorpicrin. This, like the others, is not so unfamiliar as it seems. As may be seen from its formula, CCl_3NO_2, it is formed by joining the nitric acid radical (NO_2), found in all explosives, with the main part of chloroform ($HCCl_3$). This is not quite so poisonous as phosgene, but it has the advantage that it causes nausea and vomiting. The soldier so affected is forced to take off his gas mask and then may fall victim to more toxic gases sent over simultaneously.

Chlorpicrin is a liquid and is commonly loaded in a shell or bomb with 20 per cent. of tin chloride, which produces dense white fumes that go through gas masks. It is made from picric acid (trinitrophenol), one of the best known of the high explosives, by treatment with chlorine. The chlorine is obtained, as it is in the household, from common bleaching powder, or "chloride of lime." This is mixed with water to form a cream in a steel still 18 feet high and 8 feet in diameter. A solution of calcium picrate, that is, the lime salt of picric acid, is pumped in and as the reaction begins the mixture heats up and the chlorpicrin distils over with the steam. When the distillate is condensed the chlorpicrin, being the heavier liquid, settles out under the layer of water and may be drawn off to fill the shell.

FIGHTING WITH FUMES

Much of what a student learns in the chemical laboratory he is apt to forget in later life if he does not follow it up. But there are two gases that he always remembers, chlorine and hydrogen sulfide. He is lucky if he has escaped being choked by the former or sickened by the latter. He can imagine what the effect would be if two offensive fumes could be combined without losing their offensive features. Now a combination something like this is the so-called mustard gas, which is not a gas and is not made from mustard. But it is easily gasified, and oil of mustard is about as near as Nature dare come to making such sinful stuff. It was first made by Guthrie, an Englishman, in 1860, and rediscovered by a German chemist, Victor Meyer, in 1886, but he found it so dangerous to work with that he abandoned the investigation. Nobody else cared to take it up, for nobody could see any use for it. So it remained in innocuous desuetude, a mere name in "Beilstein's Dictionary," together with the thousands of other organic compounds that have been invented and never utilized. But on July 12, 1917, the British holding the line at Ypres were besprinkled with this villainous substance. Its success was so great that the Germans henceforth made it their main reliance and soon the Allies followed suit. In one offensive of ten days the Germans are said to have used a million shells containing 2500 tons of mustard gas.

The making of so dangerous a compound on a large scale was one of the most difficult tasks set before the chemists of this and other countries, yet it was successfully solved. The raw materials are chlorine, alcohol and sulfur. The alcohol is passed with steam through a vertical iron tube filled with kaolin and heated. This converts the alcohol into a gas known as ethylene (C_2H_4). Passing a stream of chlorine gas into a tank of melted sulfur produces sulfur monochloride and this treated with the ethylene makes the "mustard." The final reaction was carried on at the Edgewood Arsenal in seven airtight tanks or "reactors," each having a capacity of 30,000 pounds. The ethylene gas being led into the tank and distributed through the liquid sulfur chloride by porous blocks or fine nozzles, the two chemicals combined to form what is officially named "di-chlor-di-ethyl-sulfide" ($ClC_2H_4SC_2H_4Cl$). This, however, is too big a mouthful, so even the chemists were glad to fall in with the commonalty and call it "mustard gas."

The effectiveness of "mustard" depends upon its persistence. It is a stable liquid, evaporating slowly and not easily decomposed. It lingers about trenches and dugouts and impregnates soil and cloth for days. Gas masks do not afford complete protection, for even if they are impenetrable they must be taken off some time and the gas lies in wait for that time. In some cases the masks were worn continuously for twelve hours after the attack, but when they were removed the soldiers were overpowered by the poison. A place may seem to be free from it but when the sun heats up the ground the liquid volatilizes and the vapor soaks through the clothing. As the men become warmed up by work their skin is blistered, especially under the armpits. The mustard acts like steam, producing burns that range from a mere reddening to serious

FIGHTING WITH FUMES

ulcerations, always painful and incapacitating, but if treated promptly in the hospital rarely causing death or permanent scars. The gas attacks the eyes, throat, nose and lungs and may lead to bronchitis or pneumonia. It was found necessary at the front to put all the clothing of the soldiers into the sterilizing ovens every night to remove all traces of mustard. General Johnson and his staff in the 77th Division were poisoned in their dugouts because they tried to alleviate the discomfort of their camp cots by bedding taken from a neighboring village that had been shelled the day before.

Of the 925 cases requiring medical attention at the Edgewood Arsenal 674 were due to mustard. During the month of August 3½ per cent. of the mustard plant force were sent to the hospital each day on the average. But the record of the Edgewood Arsenal is a striking demonstration of what can be done in the prevention of industrial accidents by the exercise of scientific prudence. In spite of the fact that from three to eleven thousand men were employed at the plant for the year 1918 and turned out some twenty thousand tons of the most poisonous gases known to man, there were only three fatalities and not a single case of blindness.

Besides the four toxic gases previously described, chlorine, phosgene, chlorpicrin and mustard, various other compounds have been and many others might be made. A list of those employed in the present war enumerates thirty, among them compounds of bromine, arsenic and cyanogen that may prove more formidable than any so far used. American chemists kept very mum during the war but occasionally one could not refrain from saying: "If the Kaiser knew what I know he would surrender unconditionally by telegraph." No doubt the science of chemical warfare is in its infancy and every foresighted power has concealed weapons of its own in reserve. One deadly compound, whose identity has not yet been disclosed, is known as "Lewisite," from Professor Lewis of Northwestern, who was manufacturing it at the rate of ten tons a day in the "Mouse Trap" stockade near Cleveland.

Throughout the history of warfare the art of defense has kept pace with the art of offense and the courage of man has never failed, no matter to what new danger he was exposed. As each new gas employed by the enemy was detected it became the business of our chemists to discover some method of absorbing or neutralizing it. Porous charcoal, best made from such dense wood as coconut shells, was packed in the respirator box together with layers of such chemicals as will catch the gases to be expected. Charcoal absorbs large quantities of any gas. Soda lime and potassium permanganate and nickel salts were among the neutralizers used.

The mask is fitted tightly about the face or over the head with rubber. The nostrils are kept closed with a clip so breathing must be done through the mouth and no air can be inhaled except that passing through the absorbent cylinder. Men within five miles of the front were required to wear the masks slung on their chests so they could be put on within six seconds. A well-made mask with a fresh box afforded almost complete immunity for a time and the soldiers learned within a few days to handle their masks adroitly. So the problem of defense against this new offensive was solved satisfactorily, while no such adequate protection against the older weapons of bayonet and shrapnel has yet been devised.

Then the problem of the offense was to catch the opponent with his mask off or to make him take it off. Here the lachrymators and the sternutators, the tear gases and the sneeze gases, came into play. Phenyl-carbylamine chloride would make the bravest soldier weep on the battlefield with the abandonment of a Greek hero. Di-phenyl-chloro-arsine would set him sneezing. The Germans alternated these with diabolical ingenuity so as to catch us unawares. Some shells gave off voluminous smoke or a vile stench without doing much harm, but by the time our men got used to these and grew careless about their masks a few shells of some extremely poisonous gas were mixed with them.

The ideal gas for belligerent purposes would be odorless, colorless and invisible, toxic even when diluted by a million parts of air, not set on fire or exploded by the detonator of the shell, not decomposed by water, not readily absorbed, stable enough to stand storage for six months and capable of being manufactured by the thousands of tons. No one gas will serve all aims. For instance, phosgene being very volatile and quickly dissipated is thrown into trenches that are soon to be taken while mustard gas being very tenacious could not be employed in such a case for the trenches could not be occupied if they were captured.

The extensive use of poison gas in warfare by all the belligerents is a vindication of the American protest at the Hague Conference against its prohibition. At the First Conference of 1899 Captain Mahan argued very sensibly that gas shells were no worse than other projectiles and might indeed prove more merciful and that it was illogical to prohibit a weapon merely because of its novelty. The British delegates voted with the Americans in opposition to the clause "the contracting parties agree to abstain from the use of projectiles the sole object of which is the diffusion of asphyxiating or deleterious gases." But both Great Britain and Germany later agreed to the provision. The use of poison gas by Germany without warning was therefore an act of treachery and a violation of her pledge, but the United States has consistently refused to bind herself to any such restriction. The facts reported by General Amos A. Fries, in command of the overseas branch of the American Chemical Warfare Service, give ample support to the American contention at The Hague:

Out of 1000 gas casualties there are from 30 to 40 fatalities, while out of 1000 high explosive casualties the number of fatalities run from 200 to 250. While exact figures are as yet not available concerning the men permanently crippled or blinded by high explosives one has only to witness the debarkation of a shipload of troops to be convinced that the number is very large. On the other hand there is, so far as known at present, not a single case of permanent disability or blindness among our troops due to gas and this in face of the fact that the Germans used relatively large quantities of this material.

In the light of these facts the prejudice against the use of gas must gradually give way; for the statement made to the effect that its use is contrary to the principles of humanity will apply with far greater force to the use of high explosives. As a matter of fact, for certain purposes toxic gas is an ideal agent. For example, it is difficult to imagine any agent more effective or more humane that may be used to render an opposing battery ineffective or to protect retreating troops.

Captain Mahan's argument at The Hague against the proposed prohibition of poison gas is so cogent and well expressed that it has been quoted in treatises on international law ever since. These reasons were, briefly:

1. That no shell emitting such gases is as yet in practical use or has undergone adequate experiment; consequently, a vote taken now would be taken in ignorance of the facts as to whether the results would be of a decisive character or whether injury in excess of that necessary to attain the end of warfare —the immediate disabling of the enemy—would be inflicted.
2. That the reproach of cruelty and perfidy, addressed against these supposed shells, was equally uttered formerly against firearms and torpedoes, both of which are now employed without scruple. Until we know the effects of such asphyxiating shells, there was no saying whether they would be more or less merciful than missiles now permitted. That it was illogical, and not demonstrably humane, to be tender about asphyxiating men with gas, when all are prepared to admit that it was allowable to blow the bottom out of an ironclad at midnight, throwing four or five hundred into the sea, to be choked by water, with scarcely the remotest chance of escape.

As Captain Mahan says, the same objection has been raised at the introduction of each new weapon of war, even though it proved to be no more cruel than the old. The modern rifle ball, swift and small and sterilized by heat, does not make so bad a wound as the ancient sword and spear, but we all remember how gunpowder was regarded by the dandies of Hotspur's time:

> And it was great pity, so it was,
> This villainous saltpeter should be digg'd
> Out of the bowels of the harmless earth
> Which many a good tall fellow had destroy'd
> So cowardly; and but for these vile guns
> He would himself have been a soldier.

The real reason for the instinctive aversion manifested against any new arm or mode of attack is that it reveals to us the intrinsic horror of war. We naturally revolt against premeditated homicide, but we have become so accustomed to the sword and latterly to the rifle that they do not shock us as they ought when we think of what they are made for. The Constitution of the United States prohibits the infliction of "cruel and unusual punishments." The two adjectives were apparently used almost synonymously, as though any "unusual" punishment were necessarily "cruel," and so indeed it strikes us. But our ingenious lawyers were able to persuade the courts that electrocution, though unknown to the Fathers and undeniably "unusual," was not unconstitutional. Dumdum bullets are rightfully ruled out because they inflict frightful and often incurable wounds, and the aim of humane warfare is to disable the enemy, not permanently to injure him.

In spite of the opposition of the American and British delegates the First Hague Conference adopted the clause, "The contracting powers agree to abstain from the use of projectiles the [sole] object of which is the diffusion of asphyxiating or deleterious gases." The word "sole" (*unique*) which appears in the original French text of The Hague convention is left out of the official English translation. This is a strange omission considering that the French and British defended their use of explosives which diffuse asphyxiating and deleterious gases on the ground that this was not the "sole" purpose of the bombs but merely an accidental effect of the nitric powder used.

The Hague Congress of 1907 placed in its rules for war: "It is expressly forbidden to employ poisons or poisonous weapons." But such attempts to rule out new and more effective means of warfare are likely to prove futile in any serious conflict and the restriction gives the advantage to the most unscrupulous side. We Americans, if ever we give our assent to such an agreement, would of course keep it, but our enemy—whoever he may be in the future—will be, as he always has been, utterly without principle and will not hesitate to employ any weapon against us. Besides, as the Germans held, chemical warfare favors the army that is most intelligent, resourceful and disciplined and the nation that stands highest in science and industry. This advantage, let us hope, will be on our side.

CHAPTER XIII

PRODUCTS OF THE ELECTRIC FURNACE

The control of man over the materials of nature has been vastly enhanced by the recent extension of the range of temperature at his command. When Fahrenheit stuck the bulb of his thermometer into a mixture of snow and salt he thought he had reached the nadir of temperature, so he scratched a mark on the tube where the mercury stood and called it zero. But we know that absolute zero, the total absence of heat, is 459 of Fahrenheit's degrees lower than his zero point. The modern scientist can get close to that lowest limit by making use of the cooling by the expansion principle. He first liquefies air under pressure and then releasing the pressure allows it to boil off. A tube of hydrogen immersed in the liquid air as it evaporates is cooled down until it can be liquefied. Then the boiling hydrogen is used to liquefy helium, and as this boils off it lowers the temperature to within three or four degrees of absolute zero.

The early metallurgist had no hotter a fire than he could make by blowing charcoal with a bellows. This was barely enough for the smelting of iron. But by the bringing of two carbon rods together, as in the electric arc light, we can get enough heat to volatilize the carbon at the tips, and this means over 7000 degrees Fahrenheit. By putting a pressure of twenty atmospheres onto the arc light we can raise it to perhaps

PRODUCTS OF ELECTRIC FURNACE

14,000 degrees, which is 3000 degrees hotter than the sun. This gives the modern man a working range of about 14,500 degrees, so it is no wonder that he can perform miracles.

When a builder wants to make an old house over into a new one he takes it apart brick by brick and stone by stone, then he puts them together in such new fashion as he likes. The electric furnace enables the chemist to take his materials apart in the same way. As the temperature rises the chemical and physical forces that hold a body together gradually weaken. First the solid loosens up and becomes a liquid, then this breaks bonds and becomes a gas. Compounds break up into their elements. The elemental molecules break up into their component atoms and finally these begin to throw off corpuscles of negative electricity eighteen hundred times smaller than the smallest atom. These electrons appear to be the building stones of the universe. No indication of any smaller units has been discovered, although we need not assume that in the electron science has delivered, what has been called, its "ultim-atom." The Greeks called the elemental particles of matter "atoms" because they esteemed them "indivisible," but now in the light of the X-ray we can witness the disintegration of the atom into electrons. All the chemical and physical properties of matter, except perhaps weight, seem to depend upon the number and movement of the negative and positive electrons and by their rearrangement one element may be transformed into another.

So the electric furnace, where the highest attainable

temperature is combined with the divisive and directive force of the current, is a magical machine for accomplishment of the metamorphoses desired by the creative chemist. A hundred years ago Davy, by dipping the poles of his battery into melted soda lye, saw forming on one of them a shining globule like quicksilver. It was the metal sodium, never before seen by man. Nowadays this process of electrolysis (electric loosening) is carried out daily by the ton at Niagara.

The reverse process, electro-synthesis (electric combining), is equally simple and even more important. By passing a strong electric current through a mixture of lime and coke the metal calcium disengages itself from the oxygen of the lime and attaches itself to the carbon. Or, to put it briefly,

$$\underset{\text{lime}}{CaO} + \underset{\text{coke}}{3C} \rightarrow \underset{\text{calcium carbide}}{CaC_2} + \underset{\text{carbon monoxide}}{CO}$$

This reaction is of peculiar importance because it bridges the gulf between the organic and inorganic worlds. It was formerly supposed that the substances found in plants and animals, mostly complex compounds of carbon, hydrogen and oxygen, could only be produced by "vital forces." If this were true it meant that chemistry was limited to the mineral kingdom and to the extraction of such carbon compounds as happened to exist ready formed in the vegetable and animal kingdoms. But fortunately this barrier to human achievement proved purely illusory. The organic field, once man had broken into it, proved easier to work in than the inorganic.

PRODUCTS OF ELECTRIC FURNACE

But it must be confessed that man is dreadfully clumsy about it yet. He takes a thousand horsepower engine and an electric furnace at several thousand degrees to get carbon into combination with hydrogen while the little green leaf in the sunshine does it quietly without getting hot about it. Evidently man is working as wastefully as when he used a thousand slaves to drag a stone to the pyramid or burned down a house to roast a pig. Not until his laboratory is as cool and calm and comfortable as the forest and the field can the chemist call himself completely successful.

But in spite of his clumsiness the chemist is actually making things that he wants and cannot get elsewhere. The calcium carbide that he manufactures from inorganic material serves as the raw material for producing all sorts of organic compounds. The electric furnace was first employed on a large scale by the Cowles Electric Smelting and Aluminum Company at Cleveland in 1885. On the dump were found certain lumps of porous gray stone which, dropped into water, gave off a gas that exploded at touch of a match with a splendid bang and flare. This gas was acetylene, and we can represent the reaction thus:

$$CaC_2 + 2H_2O \rightarrow C_2H_2 + CaO_2H_2$$
calcium *added* to water *gives* acetylene *and* slaked lime
carbide

We are all familiar with this reaction now, for it is acetylene that gives the dazzling light of the automobiles and of the automatic signal buoys of the seacoast. When burned with pure oxygen instead of air it gives the hottest of chemical flames, hotter even than the oxy-hydrogen blowpipe. For although a given weight of hydrogen will give off more heat when it burns than carbon will, yet acetylene will give off more heat than either of its elements or both of them when they are separate. This is because acetylene has stored up heat in its formation instead of giving it off as in most reactions, or to put it in chemical language, acetylene is an endothermic compound. It has required energy to bring the H and the C together, therefore it does not require energy to separate them, but, on the contrary, energy is released when they are separated. That is to say, acetylene is explosive not only when mixed with air as coal gas is but by itself. Under a suitable impulse acetylene will break up into its original carbon and hydrogen with great violence. It explodes with twice as much force without air as ordinary coal gas with air. It forms an explosive compound with copper, so it has to be kept out of contact with brass tubes and stopcocks. But compressed in steel cylinders and dissolved in acetone, it is safe and commonly used for welding and melting. It is a marvelous though not an unusual sight on city streets to see a man with blue glasses on cutting down through a steel rail with an oxy-acetylene blowpipe as easily as a carpenter saws off a board. With such a flame he can carve out a pattern in a steel plate in a way that reminds me of the days when I used to make brackets with a scroll saw out of cigar boxes. The torch will travel through a steel plate an inch or two thick at a rate of six to ten inches a minute.

The temperatures attainable with various fuels in the compound blowpipe are said to be:

Acetylene with oxygen.......... 7878° F.
Hydrogen with oxygen........ 6785° F.
Coal gas with oxygen.......... 6575° F.
Gasoline with oxygen......... 5788° F.

If we compare the formula of acetylene, C_2H_2, with that of ethylene, C_2H_4, or with ethane, C_2H_6, we see that acetylene could take on two or four more atoms. It is evidently what the chemists call an "unsaturated" compound, one that has not reached its limit of hydrogenation. It is therefore a very active and energetic compound, ready to pick up on the slightest instigation hydrogen or oxygen or chlorine or any other elements that happen to be handy. This is why it is so useful as a starting point for synthetic chemistry.

To build up from this simple substance, acetylene, the higher compounds of carbon and oxygen it is necessary to call in the aid of that mysterious agency, the catalyst. Acetylene is not always acted upon by water, as we know, for we see it bubbling up through the water when prepared from the carbide. But if to the water be added a little acid and a mercury salt, the acetylene gas will unite with the water forming a new compound, acetaldehyde. We can show the change most simply in this fashion:

$$C_2H_2 + H_2O \rightarrow C_2H_4O$$
acetylene *added to* water *forms* acetaldehyde

Acetaldehyde is not of much importance in itself, but is useful as a transition. If its vapor mixed with hydrogen is passed over finely divided nickel, serving as a catalyst, the two unite and we have alcohol, according to this reaction:

$$C_2H_4O + H_2 \rightarrow C_2H_6O$$
acetaldehyde *added to* hydrogen *forms* alcohol

Alcohol we are all familiar with—some of us too familiar, but the prohibition laws will correct that. The point to be noted is that the alcohol we have made from such unpromising materials as limestone and coal is exactly the same alcohol as is obtained by the fermentation of fruits and grains by the yeast plant as in wine and beer. It is not a substitute or imitation. It is not the wood spirits (methyl alcohol, CH_4O), produced by the destructive distillation of wood, equally serviceable as a solvent or fuel, but undrinkable and poisonous.

Now, as we all know, cider and wine when exposed to the air gradually turn into vinegar, that is, by the growth of bacteria the alcohol is oxidized to acetic acid. We can, if we like, dispense with the bacteria and speed up the process by employing a catalyst. Acetaldehyde, which is halfway between alcohol and acid, may also be easily oxidized to acetic acid. The relationship is readily seen by this:

$$C_2H_6O \rightarrow C_2H_4O \rightarrow C_2H_4O_2$$
alcohol   acetaldehyde   acetic acid

Acetic acid, familiar to us in a diluted and flavored form as vinegar, is when concentrated of great value in industry, especially as a solvent. I have already referred to its use in combination with cellulose as a "dope" for varnishing airplane canvas or making noninflammable film for motion pictures. Its combination

PRODUCTS OF ELECTRIC FURNACE

with lime, calcium acetate, when heated gives acetone, which, as may be seen from its formula (C_3H_6O) is closely related to the other compounds we have been considering, but it is neither an alcohol nor an acid. It is extensively employed as a solvent.

Acetone is not only useful for dissolving solids but it will under pressure dissolve many times its volume of gaseous acetylene. This is a convenient way of transporting and handling acetylene for lighting or welding.

If instead of simply mixing the acetone and acetylene in a solution we combine them chemically we can get isoprene, which is the mother substance of ordinary India rubber. From acetone also is made the "war rubber" of the Germans (methyl rubber), which I have mentioned in a previous chapter. The Germans had been getting about half their supply of acetone from American acetate of lime and this was of course shut off. That which was produced in Germany by the distillation of beech wood was not even enough for the high explosives needed at the front. So the Germans resorted to rotting potatoes—or rather let us say, since it sounds better—to the cultivation of *Bacillus macerans*. This particular bacillus converts the starch of the potato into two-thirds alcohol and one-third acetone. But soon potatoes got too scarce to be used up in this fashion, so the Germans turned to calcium carbide as a source of acetone and before the war ended they had a factory capable of manufacturing 2000 tons of methyl rubber a year. This shows the advantage of having several strings to a bow.

The reason why acetylene is such an active and acquisitive thing the chemist explains, or rather expresses, by picturing its structure in this shape:

$$H-C\equiv C-H$$

Now the carbon atoms are holding each other's hands because they have nothing else to do. There are no other elements around to hitch on to. But the two carbons of acetylene readily loosen up and keeping the connection between them by a single bond reach out in this fashion with their two disengaged arms and grab whatever alien atoms happen to be in the vicinity:

$$H-\overset{|}{\underset{|}{C}}-\overset{|}{\underset{|}{C}}-H$$

Carbon atoms belong to the quadrumani like the monkeys, so they are peculiarly fitted to forming chains and rings. This accounts for the variety and complexity of the carbon compounds.

So when acetylene gas mixed with other gases is passed over a catalyst, such as a heated mass of iron ore or clay (hydrates or silicates of iron or aluminum), it forms all sorts of curious combinations. In the presence of steam we may get such simple compounds as acetic acid, acetone and the like. But when three acetylene molecules join to form a ring of six carbon atoms we get compounds of the benzene series such as were described in the chapter on the coal-tar colors. If ammonia is mixed with acetylene we may get rings with the nitrogen atom in place of one of the carbons, like the pyridins and quinolins, pungent bases such as are found in opium and tobacco. Or if hydrogen sulfide is mixed with the acetylene we may get thiophenes, which have sulfur in the ring. So, starting with the simple combination of two atoms of carbon with two of hydrogen, we can get directly by this single process some of the most complicated compounds of the organic world, as well as many others not found in nature.

In the development of the electric furnace America played a pioneer part. Provost Smith of the University of Pennsylvania, who is the best authority on the history of chemistry in America, claims for Robert Hare, a Philadelphia chemist born in 1781, the honor of constructing the first electrical furnace. With this crude apparatus and with no greater electromotive force than could be attained from a voltaic pile, he converted charcoal into graphite, volatilized phosphorus from its compounds, isolated metallic calcium and synthesized calcium carbide. It is to Hare also that we owe the invention in 1801 of the oxy-hydrogen blowpipe, which nowadays is used with acetylene as well as hydrogen. With this instrument he was able to fuse strontia and volatilize platinum.

But the electrical furnace could not be used on a commercial scale until the dynamo replaced the battery as a source of electricity. The industrial development of the electrical furnace centered about the search for a cheap method of preparing aluminum. This is the metallic base of clay and therefore is common enough. But clay, as we know from its use in making porcelain, is very infusible and difficult to decompose. Sixty years ago aluminum was priced at $140 a pound, but one would have had difficulty in buying such a large quantity as a pound at any price. At international expositions a small bar of it might be seen in a case labeled "silver from clay." Mechanics were anxious to get the new metal, for it was light and untarnishable, but the metallurgists could not furnish it to them at a low enough price. In order to extract it from clay a more active metal, sodium, was essential. But sodium also was rare and expensive. In those days a professor of chemistry used to keep a little stick of it in a bottle under kerosene and once a year he whittled off a piece the size of a pea and threw it into water to show the class how it sizzled and gave off hydrogen. The way to get cheaper aluminum was, it seemed, to get cheaper sodium and Hamilton Young Castner set himself at this problem. He was a Brooklyn boy, a student of Chandler's at Columbia. You can see the bronze tablet in his honor at the entrance of Havemeyer Hall. In 1886 he produced metallic sodium by mixing caustic soda with iron and charcoal in an iron pot and heating in a gas furnace. Before this experiment sodium sold at $2 a pound; after it sodium sold at twenty cents a pound.

But although Castner had succeeded in his experiment he was defeated in his object. For while he was perfecting the sodium process for making aluminum the electrolytic process for getting aluminum directly was discovered in Oberlin. So the $250,000 plant of the "Aluminium Company Ltd." that Castner had got erected at Birmingham, England, did not make aluminum at all, but produced sodium for other purposes instead. Castner then turned his attention to the electrolytic method of producing sodium by the use of the power of Niagara Falls, electric power. Here in 1894 he succeeded in separating common salt into its com-

PRODUCTS OF ELECTRIC FURNACE

ponent elements, chlorine and sodium, by passing the electric current through brine and collecting the sodium in the mercury floor of the cell. The sodium by the action of water goes into caustic soda. Nowadays sodium and chlorine and their components are made in enormous quantities by the decomposition of salt. The United States Government in 1918 procured nearly 4,000,000 pounds of chlorine for gas warfare.

The discovery of the electrical process of making aluminum that displaced the sodium method was due to Charles M. Hall. He was the son of a Congregational minister and as a boy took a fancy to chemistry through happening upon an old textbook of that science in his father's library. He never knew who the author was, for the cover and title page had been torn off. The obstacle in the way of the electrolytic production of aluminum was, as I have said, because its compounds were so hard to melt that the current could not pass through. In 1886, when Hall was twenty-two, he solved the problem in the laboratory of Oberlin College with no other apparatus than a small crucible, a gasoline burner to heat it with and a galvanic battery to supply the electricity. He found that a Greenland mineral, known as cryolite (a double fluoride of sodium and aluminum), was readily fused and would dissolve alumina (aluminum oxide). When an electric current was passed through the melted mass the metal aluminum would collect at one of the poles.

In working out the process and defending his claims Hall used up all his own money, his brother's and his uncle's, but he won out in the end and Judge Taft held that his patent had priority over the French claim of

Hérault. On his death, a few years ago, Hall left his large fortune to his Alma Mater, Oberlin.

Two other young men from Ohio, Alfred and Eugene Cowles, with whom Hall was for a time associated, were the first to develop the wide possibilities of the electric furnace on a commercial scale. In 1885 they started the Cowles Electric Smelting and Aluminum Company at Lockport, New York, using Niagara power. The various aluminum bronzes made by absorbing the electrolyzed aluminum in copper attracted immediate attention by their beauty and usefulness in electrical work and later the company turned out other products besides aluminum, such as calcium carbide, phosphorus, and carborundum. They got carborundum as early as 1885 but miscalled it "crystallized silicon," so its introduction was left to E. A. Acheson, who was a graduate of Edison's laboratory. In 1891 he packed clay and charcoal into an iron bowl, connected it to a dynamo and stuck into the mixture an electric light carbon connected to the other pole of the dynamo. When he pulled out the rod he found its end encrusted with glittering crystals of an unknown substance. They were blue and black and iridescent, exceedingly hard and very beautiful. He sold them at first by the carat at a rate that would amount to $560 a pound. They were as well worth buying as diamond dust, but those who purchased them must have regretted it, for much finer crystals were soon on sale at ten cents a pound. The mysterious substance turned out to be a compound of carbon and silicon, the simplest possible compound, one atom of each, CSi. Acheson set up a factory at Niagara, where he made it in

From "America's Munitions"

THE CHLORPICRIN PLANT AT THE EDGEWOOD ARSENAL

From these stills, filled with a mixture of bleaching powder, lime, and picric acid, the poisonous gas, chlorpicrin, distills off. This plant produced 31 tons in one day.

Courtesy of the Metal and Thermit Corporation, N. Y.

REPAIRING THE BROKEN STERN POST OF THE U. S. S. NORTHERN PACIFIC, THE BIGGEST MARINE WELD IN THE WORLD

On the right the fractured stern post is shown. On the left it is being mended by means of thermit. Two crucibles each containing 700 pounds of the thermit mixture are seen on the sides of the vessel. From the bottom of these the melted steel flowed down to fill the fracture.

ten-ton batches. The furnace consisted simply of a brick box fifteen feet long and seven feet wide and deep, with big carbon electrodes at the ends. Between them was packed a mixture of coke to supply the carbon, sand to supply the silicon, sawdust to make the mass porous and salt to make it fusible.

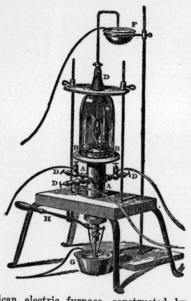

The first American electric furnace, constructed by Robert Hare of Philadelphia. From "Chemistry in America," by Edgar Fahs Smith

The substance thus produced at Niagara Falls is known as "carborundum" south of the American-Canadian boundary and as "crystolon" north of this line, as "carbolon" by another firm, and as "silicon carbide" by chemists the world over. Since it is next to the diamond in hardness it takes off metal faster than emery (aluminum oxide), using less power and wasting less heat in futile fireworks. It is used for grindstones of all sizes, including those the dentist uses on your teeth. It has revolutionized shop-practice, for articles can be ground into shape better and quicker than they can be cut. What is more, the artificial abrasives do not injure the lungs of the operatives like sandstone. The output of artificial abrasives in the United States and Canada for 1917 was:

	Tons	Value
Silicon carbide	8,323	$1,074,152
Aluminum oxide	48,463	6,969,387

A new use for carborundum was found during the war when Uncle Sam assumed the rôle of Jove as "cloud-compeller." Acting on carborundum with chlorine—also, you remember, a product of electrical dissolution—the chlorine displaces the carbon, forming silicon tetra-chloride ($SiCl_4$), a colorless liquid resembling chloroform. When this comes in contact with moist air it gives off thick, white fumes, for water decomposes it, giving a white powder (silicon hydroxide) and hydrochloric acid. If ammonia is present the acid will unite with it, giving further white fumes of the salt, ammonium chloride. So a mixture of two parts of silicon chloride with one part of dry ammonia was used in the war to produce smoke-screens for the concealment of the movements of troops, batteries and vessels or put in shells so the outlook could see where they burst and so get the range. Titanium tetra-chloride, a similar substance, proved 50 per cent. better than silicon, but phosphorus—which also we get from the electric furnace—was the most effective mistifier of all.

PRODUCTS OF ELECTRIC FURNACE

Before the introduction of the artificial abrasives fine grinding was mostly done by emery, which is an impure form of aluminum oxide found in nature. A purer form is made from the mineral bauxite by driving off its combined water. Bauxite is the ore from which is made the pure aluminum oxide used in the electric furnace for the production of metallic aluminum. Formerly we imported a large part of our bauxite from France, but when the war shut off this source we developed our domestic fields in Arkansas, Alabama and Georgia, and these are now producing half a million tons a year. Bauxite simply fused in the electric furnace makes a better abrasive than the natural emery or corundum, and it is sold for this purpose under the name of "aloxite," "alundum," "exolon," "lionite" or "coralox." When the fused bauxite is worked up with a bonding material into crucibles or muffles and baked in a kiln it forms the alundum refractory ware. Since alundum is porous and not attacked by acids it is used for filtering hot and corrosive liquids that would eat up filter-paper. Carborundum or crystolon is also made up into refractory ware for high temperature work. When the fused mass of the carborundum furnace is broken up there is found surrounding the carborundum core a similar substance though not quite so hard and infusible, known as "carborundum sand" or "siloxicon." This is mixed with fireclay and used for furnace linings.

Many new forms of refractories have come into use to meet the demands of the new high temperature work. The essentials are that it should not melt or crumble at high heat and should not expand and contract greatly under changes of temperature (low coefficient of thermal expansion). Whether it is desirable that it should heat through readily or slowly (coefficient of thermal conductivity) depends on whether it is wanted as a crucible or as a furnace lining. Lime (calcium oxide) fuses only at the highest heat of the electric furnace, but it breaks down into dust. Magnesia (magnesium oxide) is better and is most extensively employed. For every ton of steel produced five pounds of magnesite is needed. Formerly we imported 90 per cent. of our supply from Austria, but now we get it from California and Washington. In 1913 the American production of magnesite was only 9600 tons. In 1918 it was 225,000. Zirconia (zirconium oxide) is still more refractory and in spite of its greater cost zirkite is coming into use as a lining for electric furnaces.

Silicon is next to oxygen the commonest element in the world. It forms a quarter of the earth's crust, yet it is unfamiliar to most of us. That is because it is always found combined with oxygen in the form of silica as quartz crystal or sand. This used to be considered too refractory to be blown but is found to be easily manipulable at the high temperatures now at the command of the glass-blower. So the chemist rejoices in flasks that he can heat red hot in the Bunsen burner and then plunge into ice water without breaking, and the cook can bake and serve in a dish of "pyrex," which is 80 per cent. silica.

At the beginning of the twentieth century minute specimens of silicon were sold as laboratory curiosities at the price of $100 an ounce. Two years later it was turned out by the barrelful at Niagara as an accidental by-product and could not find a market at ten cents a pound. Silicon from the electric furnace appears in the form of hard, glittering metallic crystals.

An alloy of iron and silicon, ferro-silicon, made by heating a mixture of iron ore, sand and coke in the electrical furnace, is used as a deoxidizing agent in the manufacture of steel.

Since silicon has been robbed with difficulty of its oxygen it takes it on again with great avidity. This has been made use of in the making of hydrogen. A mixture of silicon (or of the ferro-silicon alloy containing 90 per cent. of silicon) with soda and slaked lime is inert, compact and can be transported to any point where hydrogen is needed, say at a battle front. Then the "hydrogenite," as the mixture is named, is ignited by a hot iron ball and goes off like thermit with the production of great heat and the evolution of a vast volume of hydrogen gas. Or the ferro-silicon may be simply burned in an atmosphere of steam in a closed tank after ignition with a pinch of gunpowder. The iron and the silicon revert to their oxides while the hydrogen of the water is set free. The French "silikol" method consists in treating silicon with a 40 per cent. solution of soda.

Another source of hydrogen originating with the electric furnace is "hydrolith," which consists of calcium hydride. Metallic calcium is prepared from lime in the electric furnace. Then pieces of the calcium are spread out in an oven heated by electricity and a current of dry hydrogen passed through. The gas is absorbed by the metal, forming the hydride (CaH_2). This is packed up in cans and when hydrogen is desired it is simply dropped into water, when it gives off the gas just as calcium carbide gives off acetylene.

This last reaction was also used in Germany for filling Zeppelins. For calcium carbide is convenient and portable and acetylene, when it is once started, as by an electric shock, decomposes spontaneously by its own internal heat into hydrogen and carbon. The latter is left as a fine, pure lampblack, suitable for printer's ink.

Napoleon, who was always on the lookout for new inventions that could be utilized for military purposes, seized immediately upon the balloon as an observation station. Within a few years after the first ascent had been made in Paris Napoleon took balloons and apparatus for generating hydrogen with him on his "archeological expedition" to Egypt in which he hoped to conquer Asia. But the British fleet in the Mediterranean put a stop to this experiment by intercepting the ship, and military aviation waited until the Great War for its full development. This caused a sudden demand for immense quantities of hydrogen and all manner of means was taken to get it. Water is easily decomposed into hydrogen and oxygen by passing an electric current through it. In various electrolytical processes hydrogen has been a wasted by-product since the balloon demand was slight and it was more bother than it was worth to collect and purify the hydrogen. Another way of getting hydrogen in quantity is by passing steam over red-hot coke. This produces the blue water-gas, which contains about 50 per cent. hydrogen, 40 per cent. carbon monoxide and the rest nitrogen and carbon dioxide. The last is removed by running the mixed gases through lime. Then the nitrogen and

PRODUCTS OF ELECTRIC FURNACE

carbon monoxide are frozen out in an air-liquefying apparatus and the hydrogen escapes to the storage tank. The liquefied carbon monoxide, allowed to regain its gaseous form, is used in an internal combustion engine to run the plant.

There are then many ways of producing hydrogen, but it is so light and bulky that it is difficult to get it where it is wanted. The American Government in the war made use of steel cylinders each holding 161 cubic feet of the gas under a pressure of 2000 pounds per square inch. Even the hydrogen used by the troops in France was shipped from America in this form. For field use the ferro-silicon and soda process was adopted. A portable generator of this type was capable of producing 10,000 cubic feet of the gas per hour.

The discovery by a Kansas chemist of natural sources of helium may make it possible to free ballooning of its great danger, for helium is non-inflammable and almost as light as hydrogen.

Other uses of hydrogen besides ballooning have already been referred to in other chapters. It is combined with nitrogen to form synthetic ammonia. It is combined with oxygen in the oxy-hydrogen blowpipe to produce heat. It is combined with vegetable and animal oils to convert them into solid fats. There is also the possibility of using it as a fuel in the internal combustion engine in place of gasoline, but for this purpose we must find some way of getting hydrogen portable or producible in a compact form.

Aluminum, like silicon, sodium and calcium, has been rescued by violence from its attachment to oxygen and like these metals it reverts with readiness to its former affinity. Dr. Goldschmidt made use of this reaction in his thermit process. Powdered aluminum is mixed with iron oxide (rust). If the mixture is heated at any point a furious struggle takes place throughout the whole mass between the iron and the aluminum as to which metal shall get the oxygen, and the aluminum always comes out ahead. The temperature runs up to some 6000 degrees Fahrenheit within thirty seconds and the freed iron, completely liquefied, runs down into the bottom of the crucible, where it may be drawn off by opening a trap door. The newly formed aluminum oxide (alumina) floats as slag on top. The applications of the thermit process are innumerable. If, for instance, it is desired to mend a broken rail or crank shaft without moving it from its place, the two ends are brought together or fixed at the proper distance apart. A crucible filled with the thermit mixture is set up above the joint and the thermit ignited with a priming of aluminum and barium peroxide to start it off. The barium peroxide having a superabundance of oxygen gives it up readily and the aluminum thus encouraged attacks the iron oxide and robs it of its oxygen. As soon as the iron is melted it is run off through the bottom of the crucible and fills the space between the rail ends, being kept from spreading by a mold of refractory material such as magnesite. The two ends of the rail are therefore joined by a section of the same size, shape, substance and strength as themselves. The same process can be used for mending a fracture or supplying a missing fragment of a steel casting of any size, such as a ship's propeller or a cogwheel.

PRODUCTS OF ELECTRIC FURNACE 257

For smaller work thermit has two rivals, the oxyacetylene torch and electric welding. The former has been described and the latter is rather out of the range of this volume, although I may mention that in the latter part of 1918 there was launched from a British shipyard the first rivetless steel vessel. In this the steel plates forming the shell, bulkheads and floors are welded instead of being fastened together by rivets. There are three methods of doing this depending upon the thickness of the plates and the sort of strain they are subject to. The plates may be overlapped and tacked together at intervals by pressing the two electrodes on opposite sides of the same point until the spot is sufficiently heated to fuse together the plates here. Or roller electrodes may be drawn slowly along the line of the desired weld, fusing the plates together continuously as they go. Or, thirdly, the plates may be butt-welded by being pushed together edge to edge without overlapping and the electric current being passed from one plate to the other heats up the joint where the conductivity is interrupted.

It will be observed that the thermit process is essentially like the ordinary blast furnace process of smelting iron and other metals except that aluminum is used instead of carbon to take the oxygen away from the metal in the ore. This has an advantage in case carbon-free metals are desired and the process is used for producing manganese, tungsten, titanium, molybdenum, vanadium and their alloys with iron and copper.

During the war thermit found a new and terrible employment, as it was used by the airmen for setting buildings on fire and exploding ammunition dumps.

The German incendiary bombs consisted of a perforated steel nose-piece, a tail to keep it falling straight and a cylindrical body which contained a tube of thermit packed around with mineral wax containing potassium perchlorate. The fuse was ignited as the missile was released and the thermit, as it heated up, melted the wax and allowed it to flow out together with the liquid iron through the holes in the nose-piece. The American incendiary bombs were of a still more malignant type. They weighed about forty pounds apiece and were charged with oil emulsion, thermit and metallic sodium. Sodium decomposes water so that if any attempt were made to put out with a hose a fire started by one of these bombs the stream of water would be instantaneously changed into a jet of blazing hydrogen.

Besides its use in combining and separating different elements the electric furnace is able to change a single element into its various forms. Carbon, for instance, is found in three very distinct forms: in hard, transparent and colorless crystals as the diamond, in black, opaque, metallic scales as graphite, and in shapeless masses and powder as charcoal, coke, lampblack, and the like. In the intense heat of the electric arc these forms are convertible one into the other according to the conditions. Since the third form is the cheapest the object is to change it into one of the other two. Graphite, plumbago or "blacklead," as it is still sometimes called, is not found in many places and more rarely found pure. The supply was not equal to the demand until Acheson worked out the process of making it by packing powdered anthracite between the elec-

PRODUCTS OF ELECTRIC FURNACE

trodes of his furnace. In this way graphite can be cheaply produced in any desired quantity and quality.

Since graphite is infusible and incombustible except at exceedingly high temperatures, it is extensively used for crucibles and electrodes. These electrodes are made in all sizes for the various forms of electric lamps and furnaces from rods one-sixteenth of an inch in diameter to bars a foot thick and six feet long. It is graphite mixed with fine clay to give it the desired degree of hardness that forms the filling of our "lead" pencils. Finely ground and flocculent graphite treated with tannin may be held in suspension in liquids and even pass through filter-paper. The mixture with water is sold under the name of "aquadag," with oil as "oildag" and with grease as "gredag," for lubrication. The smooth, slippery scales of graphite in suspension slide over each other easily and keep the bearings from rubbing against each other.

The other and more difficult metamorphosis of carbon, the transformation of charcoal into diamond, was successfully accomplished by Moissan in 1894. Henri Moissan was a toxicologist, that is to say, a Professor of Poisoning, in the Paris School of Pharmacy, who took to experimenting with the electric furnace in his leisure hours and did more to demonstrate its possibilities than any other man. With it he isolated fluorine, most active of the elements, and he prepared for the first time in their purity many of the rare metals that have since found industrial employment. He also made the carbides of the various metals, including the now common calcium carbide. Among the problems that he undertook and solved was the manufacture of artificial diamonds. He first made pure charcoal by burning sugar. This was packed with iron in the hollow of a block of lime into which extended from opposite sides the carbon rods connected to the dynamo. When the iron had melted and dissolved all the carbon it could, Moissan dumped it into water or better into melted lead or into a hole in a copper block, for this cooled it most rapidly. After a crust was formed it was left to solidify slowly. The sudden cooling of the iron on the outside subjected the carbon, which was held in solution, to intense pressure and when the bit of iron was dissolved in acid some of the carbon was found to be crystallized as diamond, although most of it was graphite. To be sure, the diamonds were hardly big enough to be seen with the naked eye, but since Moissan's aim was to make diamonds, not big diamonds, he ceased his efforts at this point.

To produce large diamonds the carbon would have to be liquefied in considerable quantity and kept in that state while it slowly crystallized. But that could only be accomplished at a temperature and pressure and duration unattainable as yet. Under ordinary atmospheric pressure carbon passes over from the solid to the gaseous phase without passing through the liquid, just as snow on a cold, clear day will evaporate without melting.

Probably some one in the future will take up the problem where Moissan dropped it and find out how to make diamonds of any size. But it is not a question that greatly interests either the scientist or the industrialist because there is not much to be learned from it and not much to be made out of it. If the inventor of a process for making cheap diamonds could keep his electric furnace secretly in his cellar and market his diamonds cautiously he might get rich out of it, but he would not dare to turn out very large stones or too many of them, for if a suspicion got around that he was making them the price would fall to almost nothing even if he did sell another one. For the high price of the diamond is purely fictitious. It is in the first place kept up by limiting the output of the natural stone by the combination of dealers and, further, the diamond is valued not for its usefulness or beauty but by its real or supposed rarity. Chesterton says: "All is gold that glitters, for the glitter is the gold." This is not so true of gold, for if gold were as cheap as nickel it would be very valuable, since we should gold-plate our machinery, our ships, our bridges and our roofs. But if diamonds were cheap they would be good for nothing except grindstones and drills. An imitation diamond made of heavy glass (paste) cannot be distinguished from the genuine gem except by an expert. It sparkles about as brilliantly, for its refractive index is nearly as high. The reason why it is not priced so highly is because the natural stone has presumably been obtained through the toil and sweat of hundreds of negroes searching in the blue ground of the Transvaal for many months. It is valued exclusively by its cost. To wear a diamond necklace is the same as hanging a certified check for $100,000 by a string around the neck.

Real values are enhanced by reduction in the cost of the price of production. Fictitious values are destroyed by it. Aluminum at twenty-five cents a pound is immensely more valuable to the world than when it is a curiosity in the chemist's cabinet and priced at $160 a pound.

So the scope of the electric furnace reaches from the costly but comparatively valueless diamond to the cheap but indispensable steel. As F. J. Tone says, if the automobile manufacturers were deprived of Niagara products, the abrasives, aluminum, acetylene for welding and high-speed tool steel, a factory now turning out five hundred cars a day would be reduced to one hundred. I have here been chiefly concerned with electricity as effecting chemical changes in combining or separating elements, but I must not omit to mention its rapidly extending use as a source of heat, as in the production and casting of steel. In 1908 there were only fifty-five tons of steel produced by the electric furnace in the United States, but by 1918 this had risen to 511,364 tons. And besides ordinary steel the electric furnace has given us alloys of iron with the once "rare metals" that have created a new science of metallurgy.

CHAPTER XIV

METALS, OLD AND NEW

The primitive metallurgist could only make use of such metals as he found free in nature, that is, such as had not been attacked and corroded by the ubiquitous oxygen. These were primarily gold or copper, though possibly some original genius may have happened upon a bit of meteoric iron and pounded it out into a sword. But when man found that the red ocher he had hitherto used only as a cosmetic could be made to yield iron by melting it with charcoal he opened a new era in civilization, though doubtless the ocher artists of that day denounced him as a utilitarian and deplored the decadence of the times.

Iron is one of the most timid of metals. It has a great disinclination to be alone. It is also one of the most altruistic of the elements. It likes almost every other element better than itself. It has an especial affection for oxygen, and, since this is in both air and water, and these are everywhere, iron is not long without a mate. The result of this union goes by various names in the mineralogical and chemical worlds, but in common language, which is quite good enough for our purpose, it is called iron rust.

Not many of us have ever seen iron, the pure metal, soft, ductile and white like silver. As soon as it is exposed to the air it veils itself with a thin film of rust and becomes black and then red. For that reason there is practically no iron in the world except what man has made. It is rarer than gold, than diamonds; we find in the earth no nuggets or crystals of it the size of the fist as we find of these. But occasionally

By courtesy *Mineral Foote-Notes*.
From Agricola's "De Re Metallica 1550." Primitive furnace for smelting iron ore.

there fall down upon us out of the clear sky great chunks of it weighing tons. These meteorites are the mavericks of the universe. We do not know where they come from or what sun or planet they belonged to. They are our only visitors from space, and if all the other spheres are like these fragments we know we are alone in the universe. For they contain rustless iron, and where iron does not rust man cannot live, nor can any other animal or any plant.

Iron rusts for the same reason that a stone rolls down hill, because it gets rid of its energy that way. All things in the universe are constantly trying to get rid of energy except man, who is always trying to get more of it. Or, on second thought, we see that man is the greatest spendthrift of all, for he wants to expend so much more energy than he has that he borrows from the winds, the streams and the coal in the rocks. He robs minerals and plants of the energy which they have stored up to spend for their own purposes, just as he robs the bee of its honey and the silk worm of its cocoon.

Man's chief business is in reversing the processes of nature. That is the way he gets his living. And one of his greatest triumphs was when he discovered how to undo iron rust and get the metal out of it. In the four thousand years since he first did this he has accomplished more than in the millions of years before. Without knowing the value of iron rust man could attain only to the culture of the Aztecs and Incas, the ancient Egyptians and Assyrians.

The prosperity of modern states is dependent on the amount of iron rust which they possess and utilize. England, United States, Germany, all nations are competing to see which can dig the most iron rust out of the ground and make out of it railroads, bridges, buildings, machinery, battleships and such other tools and toys and then let them relapse into rust again. Civilization can be measured by the amount of iron rusted per capita, or better, by the amount rescued from rust.

But we are devoting so much space to the consideration of the material aspects of iron that we are like to neglect its esthetic and ethical uses. The beauty of nature is very largely dependent upon the fact that iron rust and, in fact, all the common compounds of iron are colored. Few elements can assume so many tints. Look at the paint pot cañons of the Yellowstone. Cheap glass bottles turn out brown, green, blue, yellow or black, according to the amount and kind of iron they contain. We build a house of cream-colored brick, varied with speckled brick and adorned with terra cotta ornaments of red, yellow and green, all due to iron. Iron rusts, therefore it must be painted; but what is there better to paint it with than iron rust itself? It is cheap and durable, for it cannot rust any more than a dead man can die. And what is also of importance, it is a good, strong, clean looking, endurable color. Whenever we take a trip on the railroad and see the miles of cars, the acres of roofing and wall, the towns full of brick buildings, we rejoice that iron rust is red, not white or some less satisfying color.

We do not know why it is so. Zinc and aluminum are metals very much like iron in chemical properties, but all their salts are colorless. Why is it that the most useful of the metals forms the most beautiful compounds? Some say, Providence; some say, chance; some say nothing. But if it had not been so we would have lost most of the beauty of rocks and trees and human beings. For the leaves and the flowers would all be white, and all the men and women would look like walking corpses. Without color in the flower what would the bees and painters do? If all the grass and trees were white, it would be like winter all the year round. If we had white blood in our veins like some of the insects it would be hard lines for our poets. And what would become of our morality if we could not blush?

> "As for me, I thrill to see
>   The bloom a velvet cheek discloses!
> Made of dust! I well believe it,
>   So are lilies, so are roses."

An etiolated earth would be hardly worth living in.

The chlorophyll of the leaves and the hemoglobin of the blood are similar in constitution. Chlorophyll contains magnesium in place of iron but iron is necessary to its formation. We all know how pale a plant gets if its soil is short of iron. It is the iron in the leaves that enables the plants to store up the energy of the sunshine for their own use and ours. It is the iron in our blood that enables us to get the iron out of iron rust and make it into machines to supplement our feeble hands. Iron is for us internally the carrier of energy, just as in the form of a trolley wire or of a third rail it conveys power to the electric car. Withdraw the iron from the blood as indicated by the pallor of the cheeks, and we become weak, faint and finally die. If the amount of iron in the blood gets too small the disease germs that are always attacking us are no longer destroyed, but multiply without check and conquer us. When the iron ceases to work efficiently we are killed by the poison we ourselves generate.

Counting the number of iron-bearing corpuscles in the blood is now a common method of determining disease. It might also be useful in moral diagnosis. A microscopical and chemical laboratory attached to the courtroom would give information of more value than some of the evidence now obtained. For the anemic and the florid vices need very different treatment. An excess or a deficiency of iron in the body is liable to result in criminality. A chemical system of morals might be developed on this basis. Among the ferruginous sins would be placed murder, violence and licentiousness. Among the non-ferruginous, cowardice, sloth and lying. The former would be mostly sins of commission, the latter, sins of omission. The virtues could, of course, be similarly classified; the ferruginous virtues would include courage, self-reliance and hopefulness; the non-ferruginous, peaceableness, meekness and chastity. According to this ethical criterion the moral man would be defined as one whose conduct is better than we should expect from the per cent. of iron in his blood.

The reason why iron is able to serve this unique purpose of conveying life-giving air to all parts of the body is because it rusts so readily. Oxidation and deoxidation proceed so quietly that the tenderest cells are fed without injury. The blood changes from red to blue and *vice versa* with greater ease and rapidity than in the corresponding alternations of social status in a democracy. It is because iron is so rustable that it is so useful. The factories with big scrap-heaps of rusting machinery are making the most money. The pyramids are the most enduring structures raised by the hand of man, but they have not sheltered so many people in their forty centuries as our skyscrapers that are already rusting.

We have to carry on this eternal conflict against rust because oxygen is the most ubiquitous of the elements and iron can only escape its ardent embraces by hiding away in the center of the earth. The united elements, known to the chemist as iron oxide and to the outside world as rust, are among the commonest of compounds and their colors, yellow and red like the Spanish flag, are displayed on every mountainside. From the time of Tubal Cain man has ceaselessly labored to divorce these elements and, having once separated them, to keep them apart so that the iron may be retained in his service. But here, as usual, man is fighting against nature and his gains, as always, are only temporary. Sooner or later his vigilance is circumvented and the metal that he has extricated by the fiery furnace returns to its natural affinity. The flint arrowheads, the bronze spearpoints, the gold ornaments, the wooden idols of prehistoric man are still to be seen in our museums, but his earliest steel swords have long since crumbled into dust.

Every year the blast furnaces of the world release 72,000,000 tons of iron from its oxides and every year a large part, said to be a quarter of that amount, reverts to its primeval forms. If so, then man after five thousand years of metallurgical industry has barely got three years ahead of nature, and should he cease his efforts for a generation there would be little left to show that man had ever learned to extract iron from its ores. The old question, "What becomes of all the pins?" may be as well asked of rails, pipes and threshing machines. The end of all iron is the same. However many may be its metamorphoses while in the service of man it relapses at last into its original state of oxidation. To save a pound of iron from corrosion is then as much a benefit to the world as to produce another pound from the ore. In fact it is of much greater benefit, for it takes four pounds of coal to produce one pound of steel, so whenever a piece of iron is allowed to oxidize it means that four times as much coal must be oxidized in order to replace it. And the beds of coal will be exhausted before the beds of iron ore.

If we are ever to get ahead, if we are to gain any respite from this enormous waste of labor and natural resources, we must find ways of preventing the iron which we have obtained and fashioned into useful tools from being lost through oxidation. Now there is only one way of keeping iron and oxygen from uniting and that is to keep them apart. A very thin dividing wall will serve for the purpose, for instance, a film of oil. But ordinary oil will rub off, so it is better to cover the surface with an oil-like linseed which oxidizes to a hard elastic and adhesive coating. If with linseed oil we mix iron oxide or some other pigment we have a paint that will protect iron perfectly so long as it is unbroken. But let the paint wear off or crack so that air can get at the iron, then rust will form and spread underneath the paint on all sides. The same is true of the porcelain-like enamel with which our kitchen iron ware is nowadays coated. So long as the enamel holds it is all right but once it is broken through at any point it begins to scale off and gets into our food.

Obviously it would be better for some purposes if we could coat our iron with another and less easily oxidized metal than with such dissimilar substances as paint or porcelain. Now the nearest relative to iron is nickel, and a layer of this of any desired thickness may be easily deposited by electricity upon any surface however irregular. Nickel takes a bright polish and keeps it well, so nickel plating has become the favorite method of protection for small objects where the expense is not prohibitive. Copper plating is used for fine wires. A sheet of iron dipped in melted tin comes out coated with a thin adhesive layer of the latter metal. Such tinned plate commonly known as "tin" has become the favorite material for pans and cans. But if the tin is scratched the iron beneath rusts more rapidly than if the tin were not there, for an electrolytic action is set up and the iron, being the negative element of the couple, suffers at the expense of the tin.

With zinc it is quite the opposite. Zinc is negative toward iron, so when the two are in contact and exposed to the weather the zinc is oxidized first. A zinc plating affords the protection of a Swiss Guard, it holds out as long as possible and when broken it perishes to the last atom before it lets the oxygen get at the iron. The zinc may be applied in four different ways. (1) It may be deposited by electrolysis as in nickel plating, but the zinc coating is more apt to be porous. (2) The sheets or articles may be dipped in a bath of melted zinc. This gives us the familiar "galvanized iron," the most useful and when well done the most effective of rust preventives. Besides these older methods of applying zinc there are now two new ones. (3) One is the Schoop process by which a wire of zinc or other metal is fed into an oxyhydrogen air blast of such heat and power that it is projected as a spray of minute drops with the speed of bullets and any object subjected to the bombardment of this metallic mist receives a coating as thick as desired. The zinc spray is so fine and cool that it may be received on cloth, lace, or the bare hand. The Schoop metallizing process has recently been improved by the use of the electric current instead of the blowpipe for melting the metal. Two zinc wires connected with any electric system, preferably the direct, are fed into the "pistol." Where the wires meet an electric arc is set up and the melted zinc is sprayed out by a jet of compressed air. (4) In the Sherardizing process the articles are put into a tight drum with zinc dust and heated to 800° F. The zinc at this temperature attacks the iron and forms a series of alloys ranging from pure zinc on the top to pure iron at the bottom of the coating. Even if this cracks in part the iron is more or less protected from corrosion so long as any zinc remains. Aluminum is used similarly in the calorizing process for coating iron, copper or brass. First a surface alloy is formed by heating the metal with aluminum powder. Then the temperature is raised to a high degree so as to cause the aluminum on the surface to diffuse into the metal and afterwards it is again baked in contact with aluminum dust which puts upon it a protective plating of the pure aluminum which does not oxidize.

Another way of protecting iron ware from rusting is to rust it. This is a sort of prophylactic method like that adopted by modern medicine where inoculation with a mild culture prevents a serious attack of the disease. The action of air and water on iron forms a series of compounds and mixtures of them. Those that contain least oxygen are hard, black and magnetic like iron itself. Those that have most oxygen are red and yellow powders. By putting on a tight coating of the black oxide we can prevent or hinder the oxidation from going on into the pulverulent stage. This is done in several ways. In the Bower-Barff process the articles to be treated are put into a closed retort and a current of superheated steam passed through for twenty minutes followed by a current of producer gas (carbon monoxide), to reduce any higher oxides that may have been formed. In the Gesner process a current of gasoline vapor is used as the reducing agent. The blueing of watch hands, buckles and the like may be done by dipping them into an oxidizing bath such as melted saltpeter. But in order to afford complete protection the layer of black oxide must be thickened by repeating the process which adds to the time and expense. This causes a slight enlargement and the high temperature often warps the ware so it is not suitable for nicely adjusted parts of machinery and of course tools would lose their temper by the heat.

A new method of rust proofing which is free from these disadvantages is the phosphate process invented by Thomas Watts Coslett, an English chemist, in 1907, and developed in America by the Parker Company of Detroit. This consists simply in dipping the sheet iron or articles into a tank filled with a dilute solution of iron phosphate heated nearly to the boiling point by steam pipes. Bubbles of hydrogen stream off rapidly at first, then slower, and at the end of half an hour or longer the action ceases, and the process is complete. What has happened is that the iron has been converted into a basic iron phosphate to a depth depending upon the density of articles processed. Any one who has studied elementary qualitative analysis will remember that when he added ammonia to his "unknown" solution, iron and phosphoric acid, if present, were precipitated together, or in other words, iron phosphate is insoluble except in acids. Therefore a superficial film of such phosphate will protect the iron underneath except from acids. This film is not a coating added on the outside like paint and enamel or tin and nickel plate. It is therefore not apt to scale off and it does not increase the size of the article. No high heat is required as in the Sherardizing and Bower-Barff processes, so steel tools can be treated without losing their temper or edge.

The deposit consisting of ferrous and ferric phosphates mixed with black iron oxide may be varied in composition, texture and color. It is ordinarily a dull gray and oiling gives a soft mat black more in accordance with modern taste than the shiny nickel plating that delighted our fathers. Even the military nowadays show more quiet taste than formerly and have abandoned their glittering accoutrements.

The phosphate bath is not expensive and can be used continuously for months by adding more of the concentrated solution to keep up the strength and removing the sludge that is precipitated. Besides the iron the solution contains the phosphates of other metals such as calcium or strontium, manganese, molybdenum, or tungsten, according to the particular purpose. Since the phosphating solution does not act on nickel it may be used on articles that have been partly nickel-plated so there may be produced, for instance, a bright raised design against a dull black background. Then, too, the surface left by the Parker process is finely etched so it affords a good attachment for paint or enamel if further protection is needed. Even if the enamel does crack, the iron beneath is not so apt to rust and scale off the coating.

These, then, are some of the methods which are now being used to combat our eternal enemy, the rust that doth corrupt. All of them are useful in their several ways. No one of them is best for all purposes. The claim of "rust-proof" is no more to be taken seriously than "fire-proof." We should rather, if we were finical, have to speak of "rust-resisting" coatings as we do of "slow-burning" buildings. Nature is insidious and unceasing in her efforts to bring to ruin the achievements of mankind and we need all the weapons we can find to frustrate her destructive determination.

But it is not enough for us to make iron superficially resistant to rust from the atmosphere. We should like also to make it so that it would withstand corrosion by acids, then it could be used in place of the large and expensive platinum or porcelain evaporating pans and similar utensils employed in chemical works. This requirement also has been met in the non-corrosive forms of iron, which have come into use within the last five years. One of these, "tantiron," invented by a British metallurgist, Robert N. Lennox, in 1912, contains 15 per cent. of silicon. Similar products are known as "duriron" and "Buflokast" in America, "metilure" in France, "ileanite" in Italy and "neutraleisen" in Germany. It is a silvery-white close-grained iron, very hard and rather brittle, somewhat like cast iron but with silicon as the main additional ingredient in place of carbon. It is difficult to cut or drill but may be ground into shape by the new abrasives. It is rustproof and is not attacked by sulfuric, nitric or acetic acid, hot or cold, diluted or concentrated. It does not resist so well hydrochloric acid or sulfur dioxide or alkalies.

The value of iron lies in its versatility. It is a dozen metals in one. It can be made hard or soft, brittle or malleable, tough or weak, resistant or flexible, elastic or pliant, magnetic or non-magnetic, more or less conductive to electricity, by slight changes of composition or mere differences of treatment. No wonder that the medieval mind ascribed these mysterious transformations to witchcraft. But the modern micrometallurgist, by etching the surface of steel and photographing it, shows it up as composite as a block of granite. He is then able to pick out its component minerals, ferrite, austenite, martensite, pearlite, graphite, cementite, and to show how their abundance, shape and arrangement contribute to the strength or weakness of the specimen. The last of these constituents, cementite, is a definite chemical compound, an iron carbide, Fe_3C, containing 6.6 per cent. of carbon, so hard as to scratch glass, very brittle, and imparting these properties to hardened steel and cast iron.

With this knowledge at his disposal the iron-maker

can work with his eyes open and so regulate his melt as to cause these various constituents to crystallize out as he wants them to. Besides, he is no longer confined to the alloys of iron and carbon. He has ransacked the chemical dictionary to find new elements to add to his alloys, and some of these rarities have proved to possess great practical value. Vanadium, for instance, used to be put into a fine print paragraph in the back of the chemistry book, where the class did not get to it until the term closed. Yet if it had not been for vanadium steel we should have no Ford cars. Tungsten, too, was relegated to the rear, and if the student remembered it at all it was because it bothered him to understand why its symbol should be W instead of T. But the student of today studies his lesson in the light of a tungsten wire and relieves his mind by listening to a phonograph record played with a "tungs-tone" stylus. When I was assistant in chemistry an "analysis" of steel consisted merely in the determination of its percentage of carbon, and I used to take Saturday for it so I could have time enough to complete the combustion. Now the chemists of a steel works' laboratory may have to determine also the tungsten, chromium, vanadium, titanium, nickel, cobalt, phosphorus, molybdenum, manganese, silicon and sulfur, any or all of them, and be spry about it, because if they do not get the report out within fifteen minutes while the steel is melting in the electrical furnace the whole batch of 75 tons may go wrong. I'm glad I quit the laboratory before they got to speeding up chemists so.

The quality of the steel depends upon the presence and the relative proportions of these ingredients, and a variation of a tenth of 1 per cent. in certain of them will make a different metal out of it. For instance, the steel becomes stronger and tougher as the proportion of nickel is increased up to about 15 per cent. Raising the percentage to 25 we get an alloy that does not rust or corrode and is non-magnetic, although both its component metals, iron and nickel, are by themselves attracted by the magnet. With 36 per cent. nickel and 5 per cent. manganese we get the alloy known as "invar," because it expands and contracts very little with changes of temperature. A bar of the best form of invar will expand less than one-millionth part of its length for a rise of one degree Centigrade at ordinary atmospheric temperature. For this reason it is used in watches and measuring instruments. The alloy of iron with 46 per cent. nickel is called "platinite" because its rate of expansion and contraction is the same as platinum and glass, and so it can be used to replace the platinum wire passing through the glass of an electric light bulb.

A manganese steel of 11 to 14 per cent. is too hard to be machined. It has to be cast or ground into shape and is used for burglar-proof safes and armor plate. Chrome steel is also hard and tough and finds use in files, ball bearings and projectiles. Titanium, which the iron-maker used to regard as his implacable enemy, has been drafted into service as a deoxidizer, increasing the strength and elasticity of the steel. It is reported from France that the addition of three-tenths of 1 per cent. of zirconium to nickel steel has made it more resistant to the German perforating bullets than any steel hitherto known. The new "stainless" cutlery contains 12 to 14 per cent. of chromium.

With the introduction of harder steels came the need of tougher tools to work them. Now the virtue of a good tool steel is the same as of a good man. It must be able to get hot without losing its temper. Steel of the old-fashioned sort, as everybody knows, gets its temper by being heated to redness and suddenly cooled by quenching or plunging it into water or oil. But when the point gets heated up again, as it does by friction in a lathe, it softens and loses its cutting edge. So the necessity of keeping the tool cool limited the speed of the machine.

But about 1868 a Sheffield metallurgist, Robert F. Mushet, found that a piece of steel he was working with did not require quenching to harden it. He had it analyzed to discover the meaning of this peculiarity and learned that it contained tungsten, a rare metal unrecognized in the metallurgy of that day. Further investigation showed that steel to which tungsten and manganese or chromium had been added was tougher and retained its temper at high temperature better than ordinary carbon steel. Tools made from it could be worked up to a white heat without losing their cutting power. The new tools of this type invented by "Efficiency" Taylor at the Bethlehem Steel Works in the nineties have revolutionized shop practice the world over. A tool of the old sort could not cut at a rate faster than thirty feet a minute without overheating, but the new tungsten tools will plow through steel ten times as fast and can cut away a ton of the material in an hour. By means of these high-speed tools the United States was able to turn out five times the munitions that it could otherwise have done in the same time. On the other hand, if Germany alone had possessed the secret of the modern steels no power could have withstood her. A slight superiority in metallurgy has been the deciding factor in many a battle. Those of my readers who have had the advantages of Sunday school training will recall the case described in I Samuel 13:19–22.

By means of these new metals armor plate has been made invulnerable—except to projectiles pointed with similar material. Flying has been made possible through engines weighing no more than two pounds per horse power. The cylinders of combustion engines and the casing of cannon have been made to withstand the unprecedented pressure and corrosive action of the fiery gases evolved within. Castings are made so hard that they cannot be cut—save with tools of the same sort. In the high-speed tools now used 20 or 30 per cent. of the iron is displaced by other ingredients; for example, tungsten from 14 to 25 per cent., chromium from 2 to 7 per cent., vanadium from $\frac{1}{2}$ to $1\frac{1}{2}$ per cent., carbon from .6 to .8 per cent., with perhaps cobalt up to 4 per cent. Molybdenum or uranium may replace part of the tungsten.

Some of the newer alloys for high-speed tools contain no iron at all. That which bears the poetic name of star-stone, stellite, is composed of chromium, cobalt and tungsten in varying proportions. Stellite keeps a hard cutting edge and gets tougher as it gets hotter. It is very hard and as good for jewelry as platinum except that it is not so expensive. Cooperite, its rival, is an alloy of nickel and zirconium, stronger, lighter and cheaper than stellite.

Before the war nearly half of the world's supply of tungsten ore (wolframite) came from Burma. But although Burma had belonged to the British for a hundred years they had not developed its mineral resources and the tungsten trade was monopolized by the Germans. All the ore was shipped to Germany and the British Admiralty was content to buy from the Germans what tungsten was needed for armor plate and heavy guns. When the war broke out the British had the ore supply, but were unable at first to work it because they were not familiar with the processes. Germany, being short of tungsten, had to sneak over a little from Baltimore in the submarine *Deutschland*. In the United States before the war tungsten ore was selling at $6.50 a unit, but by the beginning of 1916 it had jumped to $85 a unit. A unit is 1 per cent. of tungsten trioxide to the ton, that is, twenty pounds. Boulder County, Colorado, and San Bernardino, California, then had mining booms, reminding one of older times. Between May and December, 1918, there was manufactured in the United States more than 45,500,000 pounds of tungsten steel containing some 8,000,000 pounds of tungsten.

If tungsten ores were more abundant and the metal more easily manipulated, it would displace steel for many purposes. It is harder than steel or even quartz. It never rusts and is insoluble in acids. Its expansion by heat is one-third that of iron. It is more than twice as heavy as iron and its melting point is twice as high.

Its electrical resistance is half that of iron and its tensile strength is a third greater than the strongest steel. It can be worked into wire .0002 of an inch in diameter, almost too thin to be seen, but as strong as copper wire ten times the size.

The tungsten wires in the electric lamps are about .03 of an inch in diameter, and they give three times the light for the same consumption of electricity as the old carbon filament. The American manufacturers of the tungsten bulb have very appropriately named their lamp "Mazda" after the light god of the Zoroastrians. To get the tungsten into wire form was a problem that long baffled the inventors of the world, for it was too refractory to be melted in mass and too brittle to be drawn. Dr. W. D. Coolidge succeeded in accomplishing the feat in 1912 by reducing the tungstic acid by hydrogen and molding the metallic powder into a bar by pressure. This is raised to a white heat in the electric furnace, taken out and rolled down, and the process repeated some fifty times, until the wire is small enough so it can be drawn at a red heat through diamond dies of successively smaller apertures.

The German method of making the lamp filaments is to squirt a mixture of tungsten powder and thorium oxide through a perforated diamond of the desired diameter. The filament so produced is drawn through a chamber heated to 2500° C. at a velocity of eight feet an hour, which crystallizes the tungsten into a continuous thread.

The first metallic filament used in the electric light on a commercial scale was made of tantalum, the metal of Tantalus. In the period 1905–1911 over 100,000,000 tantalus lamps were sold, but tungsten displaced them as soon as that metal could be drawn into wire.

A recent rival of tungsten both as a filament for lamps and hardener for steel is molybdenum. One pound of this metal will impart more resiliency to steel than three or four pounds of tungsten. The molybdenum steel, because it does not easily crack, is said to be serviceable for armor-piercing shells, gun linings, airplane struts, automobile axles and propeller shafts. In combination with its rival as a tungsten-molybdenum alloy it is capable of taking the place of the intolerably expensive platinum, for it resists corrosion when used for spark plugs and tooth plugs. European steel men have taken to molybdenum more than Americans. The salts of this metal can be used in dyeing and photography.

Calcium, magnesium and aluminum, common enough in their compounds, have only come into use as metals since the invention of the electric furnace. Now the photographer uses magnesium powder for his flashlight when he wants to take a picture of his friends inside the house, and the aviator uses it when he wants to take a picture of his enemies on the open field. The flares prepared by our Government for the war consist of a sheet iron cylinder, four feet long and six inches thick, containing a stick of magnesium attached to a tightly rolled silk parachute twenty feet in diameter when expanded. The whole weighed 32 pounds. On being dropped from the plane by pressing a button, the rush of air set spinning a pinwheel at the bottom which ignited the magnesium stick and detonated a charge of black powder sufficient to throw off the case and release the parachute. The burning flare gave off a light of 320,000 candle power lasting for ten minutes as the parachute slowly descended. This illuminated the ground on the darkest night sufficiently for the airman to aim his bombs or to take photographs.

The addition of 5 or 10 per cent. of magnesium to aluminum gives an alloy (magnalium) that is almost as light as aluminum and almost as strong as steel. An alloy of 90 per cent. aluminum and 10 per cent. calcium is lighter and harder than aluminum and more resistant to corrosion. The latest German airplane, the "Junker," was made entirely of duralumin. Even the wings were formed of corrugated sheets of this alloy instead of the usual doped cotton-cloth. Duralumin is composed of about 85 per cent. of aluminum, 5 per cent. of copper, 5 per cent. of zinc and 2 per cent. of tin.

When platinum was first discovered it was so cheap that ingots of it were gilded and sold as gold bricks to unwary purchasers. The Russian Government used it as we use nickel, for making small coins. But this is an exception to the rule that the demand creates the supply. Platinum is really a "rare metal," not merely an unfamiliar one. Nowhere except in the Urals is it found in quantity, and since it seems indispensable in chemical and electrical appliances, the price has continually gone up. Russia collapsed into chaos just when the war work made the heaviest demand for platinum, so the governments had to put a stop to its use for jewelry and photography. The "gold brick" scheme would now have to be reversed, for gold is used as a cheaper metal to "adulterate" platinum. All the mem-

METALS, OLD AND NEW

bers of the platinum family, formerly ignored, were pressed into service, palladium, rhodium, osmium, iridium, and these, alloyed with gold or silver, were employed more or less satisfactorily by the dentist, chemist and electrician as substitutes for the platinum of which they had been deprived. One of these alloys, composed of 20 per cent. palladium and 80 per cent. gold, and bearing the telescoped name of "palau" (palladium au-rum) makes very acceptable crucibles for the laboratory and only costs half as much as platinum. "Rhotanium" is a similar alloy recently introduced. The points of our gold pens are tipped with an osmium-iridium alloy. It is a pity that this family of noble metals is so restricted, for they are unsurpassed in tenacity and incorruptibility. They could be of great service to the world in war and peace. As the "Bad Child" says in his "Book of Beasts":

I shoot the hippopotamus with bullets made of platinum,
Because if I use leaden ones, his hide is sure to flatten 'em.

Along in the latter half of the last century chemists had begun to perceive certain regularities and relationships among the various elements, so they conceived the idea that some sort of a pigeon-hole scheme might be devised in which the elements could be filed away in the order of their atomic weights so that one could see just how a certain element, known or unknown, would behave from merely observing its position in the series. Mendeléef, a Russian chemist, devised the most ingenious of such systems called the "periodic law" and gave proof that there was something in his theory by predicting the properties of three metallic elements, then unknown but for which his arrangement showed three empty pigeon-holes. Sixteen years later all three of these predicted elements had been discovered, one by a Frenchman, one by a German and one by a Scandinavian, and named from patriotic impulse, gallium, germanium and scandium. This was a triumph of scientific prescience as striking as the mathematical proof of the existence of the planet Neptune by Leverrier before it had been found by the telescope.

But although Mendeléef's law told "the truth," it gradually became evident that it did not tell "the whole truth and nothing but the truth," as the lawyers put it. As usually happens in the history of science the hypothesis was found not to explain things so simply and completely as was at first assumed. The anomalies in the arrangement did not disappear on closer study, but stuck out more conspicuously. Though Mendeléef had pointed out three missing links, he had failed to make provision for a whole group of elements since discovered, the inert gases of the helium-argon group. As we now know, the scheme was built upon the false assumptions that the elements are immutable and that their atomic weights are invariable.

The elements that the chemists had most difficulty in sorting out and identifying were the heavy metals found in the "rare earths." There were about twenty of them so mixed up together and so much alike as to baffle all ordinary means of separating them. For a hundred years chemists worked over them and quarreled over them before they discovered that they had a commercial value. It was a problem as remote from practicality as any that could be conceived. The man in the street did not see why chemists should care whether there were two didymiums any more than why theologians should care whether there were two Isaiahs. But all of a sudden, in 1885, the chemical puzzle became a business proposition. The rare earths became household utensils and it made a big difference with our monthly gas bills whether the ceria and the thoria in the burner mantles were absolutely pure or contained traces of some of the other elements that were so difficult to separate.

This sudden change of venue from pure to applied science came about through a Viennese chemist, Dr. Carl Auer, later and in consequence known as Baron Auer von Welsbach. He was trying to sort out the rare earths by means of the spectroscopic method, which consists ordinarily in dipping a platinum wire into a solution of the unknown substance and holding it in a colorless gas flame. As it burns off, each element gives a characteristic color to the flame, which is seen as a series of lines when looked at through the spectroscope. But the flash of the flame from the platinum wire was too brief to be studied, so Dr. Auer hit upon the plan of soaking a thread in the liquid and putting this in the gas jet. The cotton of course burned off at once, but the earths held together and when heated gave off a brilliant white light, very much like the calcium or limelight which is produced by heating a stick of quicklime in the oxy-hydrogen flame. But these rare earths do not require any such intense heat as that, for they will glow in an ordinary gas jet.

So the Welsbach mantle burner came into use everywhere and rescued the coal gas business from the destruction threatened by the electric light. It was no longer necessary to enrich the gas with oil to make its flame luminous, for a cheaper fuel gas such as is used for a gas stove will give, with a mantle, a fine white light of much higher candle power than the ordinary gas jet. The mantles are knit in narrow cylinders on machines, cut off at suitable lengths, soaked in a solution of the salts of the rare earths and dried. Artificial silk (viscose) has been found better than cotton thread for the mantles, for it is solid, not hollow, more uniform in quality and continuous instead of being broken up into one-inch fibers. There is a great deal of difference in the quality of these mantles, as every one who has used them knows. Some that give a bright glow at first with the gas-cock only half open will soon break up or grow dull and require more gas to get any kind of a light out of them. Others will last long and grow better to the last. Slight impurities in the earths or the gas will speedily spoil the light. The best results are obtained from a mixture of 99 parts thoria and 1 part ceria. It is the ceria that gives the light, yet a little more of it will lower the luminosity.

The non-chemical reader is apt to be confused by the strange names and their varied terminations, but he need not be when he learns that new metals are given names ending in -*um*, such as sodium, cerium, thorium, and that their oxides (compounds with oxygen, the earths) are given the termination -*a*, like soda, ceria, thoria. So when he sees a name ending in -*um* let him picture to himself a metal, any metal since they mostly look alike, lead or silver, for example. And when he comes across a name ending in -*a* he may imagine a white powder like lime. Thorium, for instance, is, as its name implies, a metal named after the thunder god Thor, to whom we dedicate one day in each week, Thursday. Cerium gets its name from the Roman goddess of agriculture by way of the asteroid.

The chief sources of the material for the Welsbach burners is monazite, a glittering yellow sand composed of phosphate of cerium with some 5 per cent. of thorium. In 1916 the United States imported 2,500,000 pounds of monazite from Brazil and India, most of which used to go to Germany. In 1895 we got over a million and a half pounds from the Carolinas, but the foreign sand is richer and cheaper. The price of the salts of the rare metals fluctuates wildly. In 1895 thorium nitrate sold at $200 a pound; in 1913 it fell to $2.60, and in 1916 it rose to $8.

Since the monazite contains more cerium than thorium and the mantles made from it contain more thorium than cerium, there is a superfluity of cerium. The manufacturers give away a pound of cerium salts with every purchase of a hundred pounds of thorium salts. It annoyed Welsbach to see the cerium residues thrown away and accumulating around his mantle factory, so he set out to find some use for it. He reduced the mixed earths to a metallic form and found that it gave off a shower of sparks when scratched. An alloy of cerium with 30 or 35 per cent. of iron proved the best and was put on the market in the form of automatic lighters. A big business was soon built up in Austria on the basis of this obscure chemical element rescued from the dump-heap. The sale of the cerite lighters in France threatened to upset the finances of the republic, which derived large revenue from its monopoly of match-making, so the French Government imposed a tax upon every man who carried one. American tourists who bought these lighters in Germany used to be much annoyed at being held up on the French frontier and compelled to take out a license. During the war the cerium sparklers were much used in the trenches for lighting cigarettes, but—as those who have seen "The Better 'Ole" will know—they sometimes fail to strike fire. Auer-metal or cerium-iron alloy was used in munitions to ignite hand grenades and to blazon the flight of trailer shells. There are many other pyrophoric (light-producing) alloys, including steel, which our ancestors used with flint before matches and percussion caps were invented.

There are more than fifty metals known and not half of them have come into common use, so there is still plenty of room for the expansion of the science of metallurgy. If the reader has not forgotten his arithmetic of permutations he can calculate how many different alloys may be formed by varying the combinations and proportions of these fifty. We have seen how quickly elements formerly known only to chemists—and to some of them known only by name—have become indispensable in our daily life. Any one of those still unutilized may be found to have peculiar properties that fit it for filling a long unfelt want in modern civilization.

Who, for instance, will find a use for gallium, the metal of France? It was described in 1869 by Mendeléef in advance of its advent and has been known in person since 1875, but has not yet been set to work. It is such a remarkable metal that it must be good for something. If you saw it in a museum case on a cold day you might take it to be a piece of aluminum, but if the curator let you hold it in your hand—which he won't—it would melt and run over the floor like mercury. The melting point is 87° Fahr. It might be used in thermometers for measuring temperatures above the boiling point of mercury were it not for the peculiar fact that gallium wets glass so it sticks to the side of the tube instead of forming a clear convex curve on top like mercury.

Then there is columbium, the American metal. It is strange that an element named after Columbia should prove so impractical. Columbium is a metal closely resembling tantalum and tantalum found a use as electric light filaments. A columbium lamp should appeal to our patriotism.

The so-called "rare elements" are really abundant enough considering the earth's crust as a whole, though they are so thinly scattered that they are usually overlooked and hard to extract. But whenever one of them is found valuable it is soon found available. A systematic search generally reveals it somewhere in sufficient quantity to be worked. Who, then, will be the first to discover a use for indium, germanium, terbium, thulium, lanthanum, neodymium, scandium, samarium and others as unknown to us as tungsten was to our fathers?

As evidence of the statement that it does not matter how rare an element may be it will come into common use if it is found to be commonly useful, we may refer to radium. A good rich specimen of radium ore, pitchblende, may contain as much as one part in 4,000,000. Madame Curie, the brilliant Polish Parisian, had to work for years before she could prove to the world that such an element existed and for years afterwards before she could get the metal out. Yet now we can all afford a bit of radium to light up our watch dials in the dark. The amount needed for this is infinitesimal. If it were more it would scorch our skins, for radium is an element in eruption. The atom throws off corpuscles at intervals as a Roman candle throws off blazing balls. Some of these particles, the alpha rays, are atoms of another element, helium, charged with positive electricity and are ejected with a velocity of 18,000 miles a second. Some of them, the beta rays, are negative electrons, only about one seven-thousandth the size of the others, but are ejected with almost the speed of light, 186,000 miles a second. If one of the alpha projectiles strikes a slice of zinc sulfide it makes a splash of light big enough to be seen with a microscope, so we can now follow the flight of a single atom. The luminous watch dials consist of a coating of zinc sulfide under continual bombardment by the radium projectiles. Sir William Crookes invented this radium light apparatus and called it a "spinthariscope," which is Greek for "spark-seer."

Evidently if radium is so wasteful of its substance it cannot last forever nor could it have forever existed. The elements then are not necessarily eternal and immutable, as used to be supposed. They have a natural length of life; they are born and die and propagate, at least some of them do. Radium, for instance, is the offspring of ionium, which is the great-great-grandson of uranium, the heaviest of known elements. Putting this chemical genealogy into biblical language we might say: Uranium lived 5,000,000,000 years and begot Uranium X1, which lived 24.6 days and begot Uranium X2, which lived 69 seconds and begot Uranium 2, which lived 2,000,000 years and begot Ionium, which lived 200,000 years and begot Radium, which lived 1850 years and begot Niton, which lived 3.85 days and begot Radium A, which lived 3 minutes and begot Radium B, which lived 26.8 minutes and begot Radium C, which lived 19.5 minutes and begot Radium D, which lived 12 years and begot Radium E, which lived 5 days and begot Polonium, which lived 136 days and begot Lead.

The figures I have given are the times when half the parent substance has gone over into the next generation. It will be seen that the chemist is even more liberal in his allowance of longevity than was Moses with the patriarchs. It appears from the above that half of the radium in any given specimen will be transformed in about 2000 years. Half of what is left will disappear in the next 2000 years, half of that in the next 2000 and so on. The reader can figure out for himself when it will all be gone. He will then have the answer to the old Eleatic conundrum of when Achilles will overtake the tortoise. But we may say that after 100,000 years there would not be left any radium worth mentioning, or in other words practically all the radium now in existence is younger than the human race. The lead that is found in uranium and has presumably descended from uranium, behaves like other lead but is lighter. Its atomic weight is only 206, while ordinary lead weighs 207. It appears then that the same chemical element may have different atomic weights according to its ancestry, while on the other hand different chemical elements may have the same atomic weight. This would have seemed shocking heresy to the chemists of the last century, who prided themselves on the immutability of the elements and did not take into consideration their past life or heredity. The study of these radioactive elements has led to a new atomic theory. I suppose most of us in our youth used to imagine the atom as a little round hard ball, but now it is conceived as a sort of solar system with an electropositive nucleus acting as the sun and negative electrons revolving around it like the planets. The number of free positive electrons in the nucleus varies from one in hydrogen to 92 in uranium. This leaves room for 92 possible elements and of these all but six are more or less certainly known and definitely placed in the scheme. The atom of uranium, weighing 238 times the atom of hydrogen, is the heaviest known and therefore the ultimate limit of the elements, though it is possible that elements may be found beyond it just as the planet Neptune was discovered outside the orbit of Uranus. Considering the position of uranium and its numerous progeny as mentioned above, it is quite appropriate that this element should bear the name of the father of all the gods.

In these radioactive elements we have come upon sources of energy such as was never dreamed of in our philosophy. The most striking peculiarity of radium is that it is always a little warmer than its surroundings, no matter how warm these may be. Slowly, spontaneously and continuously, it decomposes and we know no way of hastening or of checking it. Whether it is cooled in liquefied air or heated to its melting point the change goes on just the same. An ounce of radium salt will give out enough heat in one hour to melt an ounce of ice and in the next hour will raise this water to the boiling point, and so on again and again without cessation for years, a fire without fuel, a realization of the philosopher's lamp that the alchemists sought in vain. The total energy so emitted is millions of times greater than that produced by any chemical combination such as the union of oxygen and hydrogen to form water. From the heavy white salt there is continually rising a faint fire-mist like the will-o'-the-wisp over a swamp. This gas is known as the emanation or niton, "the shining one." A pound of niton would give off energy at the rate of 23,000 horsepower; fine stuff to run a steamer, one would think, but we must remember that it does not last. By the sixth day the power would have fallen off by half. Besides, no one would dare to serve as engineer, for the radiation will rot away the flesh of a living man who comes near it, causing gnawing ulcers or curing them. It will not only break down the complex and delicate molecules of organic matter but will attack the atom itself, changing, it is believed, one element into another, again the fulfilment of a dream of the alchemists. And its rays, unseen and unfelt by us, are yet strong enough to penetrate an armorplate and photograph what is behind it.

But radium is not the most mysterious of the elements but the least so. It is giving out the secret that the other elements have kept. It suggests to us that all the other elements in proportion to their weight have concealed within them similar stores of energy. Astronomers have long dazzled our imaginations by calculating the horsepower of the world, making us feel cheap in talking about our steam engines and dynamos when a minutest fraction of the waste dynamic energy of the solar system would make us all as rich as millionaires. But the heavenly bodies are too big for us to utilize in this practical fashion.

And now the chemists have become as exasperating as the astronomers, for they give us a glimpse of incalculable wealth in the meanest substance. For wealth is measured by the available energy of the world, and if a few ounces of anything would drive an engine or manufacture nitrogenous fertilizer from the air all our troubles would be over. Kipling in his sketch, "With the Night Mail," and Wells in his novel, "The World Set Free," stretched their imaginations in trying to tell us what it would mean to have command of this power, but they are a little hazy in their descriptions of the machinery by which it is utilized. The atom is as much beyond our reach as the moon. We cannot rob its vault of the treasure.

READING REFERENCES

The foregoing pages will not have achieved their aim unless their readers have become sufficiently interested in the developments of industrial chemistry to desire to pursue the subject further in some of its branches. Assuming such interest has been aroused, I am giving below a few references to books and articles which may serve to set the reader upon the right track for additional information. To follow the rapid progress of applied science it is necessary to read continuously such periodicals as the *Journal of Industrial and Engineering Chemistry* (New York), *Metallurgical and Chemical Engineering* (New York), *Journal of the Society of Chemical Industry* (London), *Chemical Abstracts* (published by the American Chemical Society, Easton, Pa.), and the various journals devoted to special trades. The reader may need to be reminded that the United States Government publishes for free distribution or at low price annual volumes or special reports dealing with science and industry. Among these may be mentioned "Yearbook of the Department of Agriculture"; "Mineral Resources of the United States," published by the United States Geological Survey in two annual volumes, Vol. I on the metals and Vol. II on the non-metals; the "Annual Report of the Smithsonian Institution," containing selected articles on pure and applied science; the daily "Commerce Reports" and special bulletins of Department of Commerce. Write for lists of publications of these departments.

The following books on industrial chemistry in general are recommended for reading and reference: "The Chemistry of Commerce" and "Some Chemical Problems of To-Day" by

Robert Kennedy Duncan (Harpers, N. Y.), "Modern Chemistry and Its Wonders" by Martin (Van Nostrand), "Chemical Discovery and Invention in the Twentieth Century" by Sir William A. Tilden (Dutton, N. Y.), "Discoveries and Inventions of the Twentieth Century" by Edward Cressy (Dutton), "Industrial Chemistry" by Allen Rogers (Van Nostrand).

"Everyman's Chemistry" by Ellwood Hendrick (Harpers, Modern Science Series) is written in a lively style and assumes no previous knowledge of chemistry from the reader. The chapters on cellulose, gums, sugars and oils are particularly interesting. "Chemistry of Familiar Things" by S. S. Sadtler (Lippincott) is both comprehensive and comprehensible.

The following are intended for young readers but are not to be despised by their elders who may wish to start in on an easy up-grade: "Chemistry of Common Things" (Allyn & Bacon, Boston) is a popular high school text-book but differing from most text-books in being readable and attractive. Its descriptions of industrial processes are brief but clear. The "Achievements of Chemical Science" by James C. Philip (Macmillan) is a handy little book, easy reading for pupils. "Introduction to the Study of Science" by W. P. Smith and E. G. Jewett (Macmillan) touches upon chemical topics in a simple way.

On the history of commerce and the effect of inventions on society the following titles may be suggested: "Outlines of Industrial History" by E. Cressy (Macmillan); "The Origin of Invention," a study of primitive industry, by O. T. Mason (Scribner); "The Romance of Commerce" by Gordon Selbridge (Lane); "Industrial and Commercial Geography" or "Commerce and Industry" by J. Russell Smith (Holt); "Handbook of Commercial Geography" by G. G. Chisholm (Longmans).

The newer theories of chemistry and the constitution of the

READING REFERENCES

atom are explained in "The Realities of Modern Science" by John Mills (Macmillan), and "The Electron" by R. A. Millikan (University of Chicago Press), but both require a knowledge of mathematics. The little book on "Matter and Energy" by Frederick Soddy (Holt) is better adapted to the general reader. The most recent text-book is the "Introduction to General Chemistry" by H. N. McCoy and E. M. Terry. (Chicago, 1919.)

CHAPTER II

The reader who may be interested in following up this subject will find references to all the literature in the summary by Helen R. Hosmer, of the Research Laboratory of the General Electric Company, in the *Journal of Industrial and Engineering Chemistry*, New York, for April, 1917. Bucher's paper may be found in the same journal for March, and the issue for September contains a full report of the action of U. S. Government and a comparison of the various processes. Send fifteen cents to the U. S. Department of Commerce (or to the nearest custom house) for Bulletin No. 52, Special Agents Series on "Utilization of Atmospheric Nitrogen" by T. H. Norton. The Smithsonian Institution of Washington has issued a pamphlet on "Sources of Nitrogen Compounds in the United States." In the 1913 report of the Smithsonian Institution there are two fine articles on this subject: "The Manufacture of Nitrates from the Atmosphere" and "The Distribution of Mankind," which discusses Sir William Crookes' prediction of the exhaustion of wheat land. The D. Van Nostrand Co., New York, publishes a monograph on "Fixation of Atmospheric Nitrogen" by J. Knox, also "TNT and Other Nitrotoluenes" by G. C. Smith. The American Cyanamid Company, New York, gives out some attractive literature on their process.

"American Munitions 1917–1918," the report of Benedict Crowell, Director of Munitions, to the Secretary of War, gives a fully illustrated account of the manufacture of arms, explosives and toxic gases. Our war experience in the "Oxidation of Ammonia" is told by C. L. Parsons in *Journal of Industrial and Engineering Chemistry*, June, 1919, and various other articles on the government munition work appeared in the same journal in the first half of 1919. "The Muscle Shoals Nitrate Plant" in *Chemical and Metallurgical Engineering*, January, 1919.

CHAPTER III

The Department of Agriculture or your congressman will send you literature on the production and use of fertilizers. From your state agricultural experiment station you can procure information as to local needs and products. Consult the articles on potash salts and phosphate rock in the latest volume of "Mineral Resources of the United States," Part II Non-Metals (published free by the U. S. Geological Survey). Also consult the latest Yearbook of the Department of Agriculture. For self-instruction, problems and experiments get "Extension Course in Soils," Bulletin No. 355, U. S. Dept. of Agric. A list of all government publications on "Soil and Fertilizers" is sent free by Superintendent of Documents, Washington. The *Journal of Industrial and Engineering Chemistry* for July, 1917, publishes an article by W. C. Ebaugh on "Potash and a World Emergency," and various articles on American sources of potash appeared in the same *Journal* October, 1918, and February, 1918. Bulletin 102, Part 2, of the United States National Museum contains an interpretation of the fertilizer situation in 1917 by J. E. Poque. On new potash deposits in Alsace and elsewhere see *Scientific American Supplement*, September 14, 1918.

CHAPTER IV

Send ten cents to the Department of Commerce, Washington, for "Dyestuffs for American Textile and Other Industries," by Thomas H. Norton, Special Agents' Series, No. 96. A more technical bulletin by the same author is "Artificial Dyestuffs Used in the United States," Special Agents' Series, No. 121, thirty cents. "Dyestuff Situation in U. S.," Special Agents' Series, No. 111, five cents. "Coal-Tar Products," by H. G. Porter, Technical Paper 89, Bureau of Mines, Department of the Interior, five cents. "Wealth in Waste," by Waldemar Kaempfert, *McClure's*, April, 1917. "The Evolution of Artificial Dyestuffs," by Thomas H. Norton, *Scientific American*, July 21, 1917. "Germany's Commercial Preparedness for Peace," by James Armstrong, *Scientific American*, January 29, 1916. "The Conquest of Commerce" and "American Made," by Edwin E. Slosson in *The Independent* of September 6 and October 11, 1915. The H. Koppers Company, Pittsburgh, give out an illustrated pamphlet on their "By-Product Coke and Gas Ovens." The addresses delivered during the war on "The Aniline Color, Dyestuff and Chemical Conditions," by I. F. Stone, president of the National Aniline and Chemical Company, have been collected in a volume by the author. For "Dyestuffs as Medicinal Agents" by G. Heyl, see *Color Trade Journal*, vol. 4, p. 73, 1919. "The Chemistry of Synthetic Drugs" by Percy May, and "Color in Relation to Chemical Constitution" by E. R. Watson are published in Longmans' "Monographs on Industrial Chemistry." "Enemy Property in the United States" by A. Mitchell Palmer in *Saturday Evening Post*, July 19, 1919, tells of how Germany monopolized chemical industry. "The Carbonization of Coal" by V. B. Lewis (Van Nostrand, 1912). "Research in the Tar Dye Industry" by B. C. Hesse in *Journal of Industrial and Engineering Chemistry*, September, 1916.

Kekulé tells how he discovered the constitution of benzene in the *Berichte der Deutschen chemischen Gesellschaft*, V. XXIII, I, p. 1306. I have quoted it with some other instances of dream discoveries in *The Independent* of Jan. 26,

1918. Even this innocent scientific vision has not escaped the foul touch of the Freudians. Dr. Alfred Robitsek in "Symbolisches Denken in der chemischen Forschung," *Imago*, V. I, p. 83, has deduced from it that Kekulé was morally guilty of the crime of Œdipus as well as minor misdemeanors.

CHAPTER V

Read up on the methods of extracting perfumes from flowers in any encyclopedia or in Duncan's "Chemistry of Commerce" or Tilden's "Chemical Discovery in the Twentieth Century" or Rogers' "Industrial Chemistry."

The pamphlet containing a synopsis of the lectures by the late Alois von Isakovics on "Synthetic Perfumes and Flavors," published by the Synfleur Scientific Laboratories, Monticello, New York, is immensely interesting. Van Dyk & Co., New York, issue a pamphlet on the composition of oil of rose. Gildemeister's "The Volatile Oils" is excellent on the history of the subject. Walter's "Manual for the Essence Industry" (Wiley) gives methods and recipes. Parry's "Chemistry of Essential Oils and Artificial Perfumes," 1918 edition. "Chemistry and Odoriferous Bodies Since 1914" by G. Satie in *Chemie et Industrie*, vol. II, p. 271, 393. "Odor and Chemical Constitution," *Chemical Abstracts*, 1917, p. 3171, and *Journal of Society for Chemical Industry*, v. 36, p. 942.

CHAPTER VI

The bulletin on "By-Products of the Lumber Industry" by H. K. Benson (published by Department of Commerce, Washington, 10 cents) contains a description of paper-making and wood distillation. There is a good article on cellulose products by H. S. Mork in *Journal of the Franklin Institute*, September, 1917, and in *Paper*, September 26, 1917. The Government Forest Products Laboratory at Madison, Wisconsin, publishes technical papers on distillation of wood, etc. The Forest Service of the U. S. Department of Agriculture is the chief source of information on forestry. The standard authority is Cross and Bevans' "Cellulose." For the acetates see the eighth volume of Worden's "Technology of the Cellulose Esters."

CHAPTER VII

The speeches made when Hyatt was awarded the Perkin medal by the American Chemical Society for the discovery of celluloid may be found in the *Journal of the Society of Chemical Industry* for 1914, p. 225. In 1916 Baekeland received the same medal, and the proceedings are reported in the same *Journal*, v. 35, p. 285.

A comprehensive technical paper with bibliography on "Synthetic Resins" by L. V. Redman appeared in the *Journal of Industrial and Engineering Chemistry*, January, 1914. The controversy over patent rights may be followed in the same *Journal*, v. 8 (1915), p. 1171, and v. 9 (1916), p. 207. The "Effects of Heat on Celluloid" have been examined by the Bureau of Standards, Washington (Technological Paper No. 98), abstract in *Scientific American Supplement*, June 29, 1918.

For casein see Tague's article in Rogers' "Industrial Chemistry" (Van Nostrand). See also Worden's "Nitrocellulose Industry" and "Technology of the Cellulose Esters" (Van Nostrand); Hodgson's "Celluloid" and Cross and Bevan's "Cellulose."

For references to recent research and new patent specifications on artificial plastics, resins, rubber, leather, wood, etc., see the current numbers of *Chemical Abstracts* (Easton, Pa.) and such journals as the *India Rubber Journal, Paper, Textile World, Leather World* and *Journal of American Leather Chemical Association*.

The General Bakelite Company, New York, the Redmanol Products Company, Chicago, the Condensite Company, Bloomfield, N. J., the Arlington Company, New York (handling pyralin), give out advertising literature regarding their respective products.

CHAPTER VIII

Sir William Tilden's "Chemical Discovery and Invention in the Twentieth Century" (E. P. Dutton & Co.) contains a readable chapter on rubber with references to his own discovery. The "Wonder Book of Rubber," issued by the B. F. Goodrich Rubber Company, Akron, Ohio, gives an interesting account of their industry. Iles: "Leading American Inventors" (Henry Holt & Co.) contains a life of Goodyear, the discoverer of vulcanization. Potts: "Chemistry of the Rubber Industry, 1912." The Rubber Industry: Report of the International Rubber Congress, 1914. Pond: "Review of Pioneer Work in Rubber Synthesis" in *Journal of the American Chemical Society*, 1914. King: "Synthetic Rubber" in *Metallurgical and Chemical Engineering*, May 1, 1917. Castellan: "L'Industrie caoutchoucière," doctor's thesis, University of Paris, 1915. The *India Rubber World*, New York, all numbers, especially "What I Saw in the Philippines," by the Editor, 1917. Pearson: "Production of Guayule Rubber," *Commerce Reports*, 1918, and *India Rubber World*, 1919. "Historical Sketch of Chemistry of Rubber" by S. C. Bradford in *Science Progress*, v. II, p. 1.

CHAPTER IX

"The Cane Sugar Industry" (Bulletin No. 53, Miscellaneous Series, Department of Commerce, 50 cents) gives agricultural and manufacturing costs in Hawaii, Porto Rico, Louisiana and Cuba.

"Sugar and Its Value as Food," by Mary Hinman Abel. (Farmer's Bulletin No. 535, Department of Agriculture, free.)

"Production of Sugar in the United States and Foreign

Countries," by Perry Elliott. (Department of Agriculture, 10 cents.)

"Conditions in the Sugar Market January to October, 1917," a pamphlet published by the American Sugar Refining Company, 117 Wall Street, New York, gives an admirable survey of the present situation as seen by the refiners.

"Cuban Cane Sugar," by Robert Wiles, 1916 (Indianapolis: Bobbs-Merrill Co., 75 cents), an attractive little book in simple language.

"The World's Cane Sugar Industry, Past and Present," by H. C. P. Geering.

"The Story of Sugar," by Prof. G. T. Surface of Yale (Appleton, 1910). A very interesting and reliable book.

The "Digestibility of Glucose" is discussed in *Journal of Industrial and Engineering Chemistry*, August, 1917. "Utilization of Beet Molasses" in *Metallurgical and Chemical Engineering*, April 5, 1917.

CHAPTER X

"Maize," by Edward Alber (Bulletin of the Pan-American Union, January, 1915).

"Glucose," by Geo. W. Rolfe (*Scientific American Supplement*, May 15 or November 6, 1915, and in Roger's "Industrial Chemistry").

On making ethyl alcohol from wood, see Bulletin No. 110, Special Agents' Series, Department of Commerce (10 cents), and an article by F. W. Kressmann in *Metallurgical and Chemical Engineering*, July 15, 1916. On the manufacture and uses of industrial alcohol the Department of Agriculture has issued for free distribution Farmer's Bulletin 269 and 424, and Department Bulletin 182.

On the "Utilization of Corn Cobs," see *Journal of Industrial and Engineering Chemistry*, Nov., 1918. For John Winthrop's experiment, see the same *Journal*, Jan., 1919.

CHAPTER XI

President Scherer's "Cotton as a World Power" (Stokes, 1916) is a fascinating volume that combines the history, science and politics of the plant and does not ignore the poetry and legend.

In the Yearbook of the Department of Agriculture for 1916 will be found an interesting article by H. S. Bailey on "Some American Vegetable Oils" (sold separate for five cents), also "The Peanut: A Great American Food" by same author in the Yearbook of 1917. "The Soy Bean Industry" is discussed in the same volume. See also: Thompson's "Cottonseed Products and Their Competitors in Northern Europe" (Part I, Cake and Meal; Part II, Edible Oils. Department of Commerce, 10 cents each). "Production and Conservation of Fats and Oils in the United States" (Bulletin No. 769, 1919, U. S. Dept. of Agriculture). "Cottonseed Meal for Feeding Cattle" (U. S. Department of Agriculture, Farmer's Bulletin 655, free). "Cottonseed Industry in Foreign Countries," by T. H. Norton, 1915 (Department of Commerce, 10 cents). "Cottonseed Products" in *Journal of the Society of Chemical Industry*, July 16, 1917, and Baskerville's article in the same journal (1915, vol. 7, p. 277). Dunstan's "Oil Seeds and Feeding Cakes," a volume on British problems since the war. Ellis's "The Hydrogenation of Oils" (Van Nostrand, 1914). Copeland's "The Coconut" (Macmillan). Barrett's "The Philippine Coconut Industry" (Bulletin No. 25, Philippine Bureau of Agriculture). "Coconuts, the Consols of the East" by Smith and Pope (London). "All About Coconuts" by Belfort and Hoyer (London). Numerous articles on copra and other oils appear in *U. S. Commerce Reports* and *Philippine Journal of Science*. "The World Wide Search for Oils" in *The Americas* (National City Bank, N. Y.). "Modern Margarine Technology" by W. Clayton in *Journal Society of Chemical Industry*, Dec. 5, 1917; also see *Scientific*

American Supplement, Sept. 21, 1918. A court decision on the patent rights of hydrogenation is given in *Journal of Industrial and Engineering Chemistry* for December, 1917. The standard work on the whole subject is Lewkowitsch's "Chemical Technology of Oils, Fats and Waxes" (3 vols., Macmillan, 1915).

CHAPTER XII

A full account of the development of the American Warfare Service has been published in the *Journal of Industrial and Engineering Chemistry* in the monthly issues from January to August, 1919, and an article on the British service in the issue of April, 1918. See also Crowell's Report on "America's Munitions," published by War Department. *Scientific American*, March 29, 1919, contains several articles. A. Russell Bond's "Inventions of the Great War" (Century) contains chapters on poison gas and explosives.

Lieutenant Colonel S. J. M. Auld, Chief Gas Officer of Sir Julian Byng's army and a member of the British Military Mission to the United States, has published a volume on "Gas and Flame in Modern Warfare" (George H. Doran Co.).

CHAPTER XIII

See chapter in Cressy's "Discoveries and Inventions of Twentieth Century." "Oxy-Acetylene Welders," Bulletin No. 11, Federal Board of Vocational Education, Washington, June, 1918, gives practical directions for welding. *Reactions*, a quarterly published by Goldschmidt Thermit Company, N. Y., reports latest achievements of aluminothermics. Provost Smith's "Chemistry in America" (Appleton) tells of the experiments of Robert Hare and other pioneers. "Applications of Electrolysis in Chemical Industry" by A. F. Hall (Longmans). For recent work on artificial diamonds see *Scientific American Supplement*, Dec. 8, 1917, and August 24, 1918. On acetylene see "A Storehouse of Sleeping Energy" by J. M. Morehead in *Scientific American*, January 27, 1917.

CHAPTER XIV

Spring's "Non-Technical Talks on Iron and Steel" (Stokes) is a model of popular science writing, clear, comprehensive and abundantly illustrated. Tilden's "Chemical Discovery in the Twentieth Century" must here again be referred to. The Encyclopedia Britannica is convenient for reference on the various metals mentioned; see the article on "Lighting" for the Welsbach burner. The annual "Mineral Resources of the United States, Part I," contains articles on the newer metals by Frank W. Hess; see "Tungsten" in the volume for 1914, also Bulletin No. 652, U. S. Geological Survey, by same author. *Foote-Notes*, the house organ of the Foote Mineral Company, Philadelphia, gives information on the rare elements. Interesting advertising literature may be obtained from the Titantium Alloy Manufacturing Company, Niagara Falls, N. Y.; Duriron Castings Company, Dayton, O.; Buffalo Foundry and Machine Company, Buffalo, N. Y., manufacturers of "Buflokast" acid-proof apparatus, and similar concerns. The following additional references may be useful: Stellite alloys in *Jour. Ind. & Eng. Chem.*, v. 9, p. 974; Rossi's work on titanium in same journal, Feb., 1918; Welsbach mantles in *Journal Franklin Institute*, v. 14, p. 401, 585; pure alloys in *Trans. Amer. Electro-Chemical Society*, v. 32, p. 269; molybdenum in *Engineering*, 1917, or *Scientific American Supplement*, Oct. 20, 1917; acid-resisting iron in *Sc. Amer. Sup.*, May 31, 1919; ferro-alloys in *Jour. Ind. & Eng. Chem.*, v. 10, p. 831; influence of vanadium, etc., on iron, in *Met. Chem. Eng.*, v. 15, p. 530; tungsten in *Engineering*, v. 104, p. 214.

INDEX

Abrasives, 249-251
Acetanilid, 87
Acetone, 125, 154, 243, 245
Acetylene, 30, 154, 240-245, 257, 307, 308
Acheson, 249
Air, liquefied, 33
Alcohol, ethyl, 101, 102, 127, 174, 190-194, 242-244, 305
  methyl, 101, 102, 127, 191
Aluminum, 31, 246-248, 255, 272, 284
Ammonia, 27, 29, 31, 33, 56, 64, 250
American dye industry, 82
Aniline dyes, 60-92
Antiseptics, 86, 87
Argon, 16
Art and nature, 8, 9, 170, 173
Artificial silk, 116, 118, 119
Aspirin, 84
Atomic theory, 293-296, 299
Aylesworth, 140

Baekeland, 137
Baeyer, Adolf von, 77
Bakelite, 138, 303
Balata, 159
Bauxite, 31
Beet sugar, 165, 169, 305
Benzene formula, 67, 301, 101
Berkeley, 61
Berthelot, 7, 94
Birkeland-Eyde process, 26
Bucher process, 32
Butter, 201, 208

Calcium, 246, 253
Calcium carbide, 30, 339
Camphor, 100, 131
Cane sugar, 164, 167, 177, 180, 305
Carbolic acid, 18, 64, 84, 101, 102, 137
Carborundum, 249-251
Caro and Franke process, 30

Casein, 142
Castner, 246
Catalyst, 28, 204
Celluloid, 128-135, 302
Cellulose, 110-127, 129, 137, 302
Cellulose acetate, 118, 120, 302
Cerium, 288-290
Chemical warfare, 218-235, 307
Chlorin, 224, 226, 250
Chlorophyll, 267
Chlorpicrin, 224, 226
Chromicum, 278, 280
Coal, distillation of, 60, 64, 70, 84, 301
Coal tar colors, 60-92
Cochineal, 79
Coconut oil, 203, 211-215, 306
Collodion, 117, 123, 130
Cologne, eau de, 107
Copra, 203, 211-215, 306
Corn oil, 183, 305
Cotton, 112, 120, 129, 197
Cocain, 88
Condensite, 141
Cordite, 18, 19
Corn products, 181-195, 305
Coslett process, 273
Cottonseed oil, 201
Cowles, 248
Creative chemistry, 7
Crookes, Sir William, 292, 290
Curie, Madame, 292
Cyanamid, 30, 35, 299
Cyanides, 32

Diamond, 259-261, 308
Doyle, Sir Arthur Conan, 221
Drugs, synthetic, 6, 84, 301
Duisberg, 151
Dyestuffs, 60-92

Edison, 84, 141
Ehrlich, 86, 87
Electric furnace, 236-262, 307

Fats, 196-217, 306

INDEX

Fertilizers, 37, 41, 43, 46, 300
Flavors, synthetic, 93-109
Food, synthetic, 94
Formaldehyde, 136, 142
Fruit flavors, synthetic, 99, 101

Galalith, 142
Gas masks, 223, 226, 230, 231
Gerhardt, 6, 7
Glucose, 137, 184-189, 194, 305
Glycerin, 194, 203
Goldschmidt, 256
Goodyear, 161
Graphite, 258
Guayule, 159, 304
Guncotton, 17, 117, 125, 130
Gunpowder, 14, 15, 22, 234
Gutta percha, 159

Haber process, 27, 28
Hall, C. H., 247
Hare, Robert, 237, 245, 307
Harries, 149
Helium, 236
Hesse, 70, 72, 90
Hofmann, 72, 80
Huxley, 10
Hyatt, 128, 129, 303
Hydrogen, 253-255
Hydrogenation of oils, 202-205, 306

Indigo, 76, 79
Iron, 236, 253, 262-270, 308
Isoprene, 136, 146, 149, 150, 154

Kelp products, 53, 142
Kekulé's dream, 66, 301

Lard substitutes, 209
Lavoisier, 6
Leather substitutes, 124
Leucite, 53
Liebig, 38
Linseed oil, 202, 205, 270

Magnesium, 283
Maize products, 181-196, 305
Manganese, 278
Margarin, 207-212, 307
Mauve, discovery of, 74
Mendeleef, 285, 291

Mercerized cotton, 115
Moissan, 259
Molybdenum, 283, 308
Munition manufacture in U. S., 33, 224, 299, 307
Mushet, 279
Musk, synthetic, 96, 97, 106
Mustard gas, 224, 227-229

Naphthalene, 4, 142, 154
Nature and art, 8-13, 118, 122, 133
Nitrates, Chilean, 22, 24, 30, 36
Nitric acid derivatives, 20
Nitrocellulose, 17, 117
Nitrogen, in explosives, 14, 16, 117, 299
fixation, 24, 25, 29, 299
Nitro-glycerin, 18, 117, 214
Nobel, 18, 117

Oils, 196-217, 306
Oleomargarin, 207-212, 307
Orange blossoms, 99, 100
Osmium, 28
Ostwald, 29, 55
Oxy-hydrogen blowpipe, 246

Paper, 111, 132
Parker process, 273
Peanut oil, 206, 211, 214, 306
Perfumery, Art of, 103-108
Perfumes, synthetic, 93-109, 302
Perkin, W. H., 148
Perkin, Sir William, 72, 80, 102
Pharmaceutical chemistry, 6, 85-88
Phenol, 18, 64, 84, 101, 102, 137
Phonograph records, 84, 141
Phosphates, 56-59
Phosgene, 224, 225
Photographic developers, 88
Picric acid, 18, 84, 85, 226
Platinum, 28, 278, 280, 284, 286
Plastics, synthetic, 128-143
Pneumatic tires, 162
Poisonous gases in warfare, 218, 235, 307
Potash, 37, 45-56, 300
Priestley, 150, 160
Purple, royal, 75, 79
Pyralin, 132, 133

INDEX

Pyrophoric alloys, 290
Pyroxylin, 17, 117, 125, 130

Radium, 291, 295
Rare earths, 286–288, 308
Redmanol, 140
Remsen, Ira, 178
Refractories, 251–252
Resins, synthetic, 135–143
Rose perfume, 93, 96, 97, 99, 105
Rubber, natural, 155–161, 304
 synthetic, 136, 145–163, 304
Rumford, Count, 166
Rust, protection from, 262–275

Saccharin, 178, 179
Salicylic acid, 88, 101
Saltpeter, Chilean, 22, 30, 36, 42
Smith, Provost, 237, 245, 307
Smokeless powder, 15
Sodium, 148, 238, 247
Soil chemistry, 38, 39
Soy bean, 142, 211, 217, 306
Starch, 137, 184, 189, 190
Stassfort salts, 47, 49, 55
Stellites, 280, 308
Sugar, 164–180, 304
Sulfuric acid, 57

Tantalum, 282
Terpenes, 100, 154
Textile industry, 5, 112, 121, 300
Thermit, 256
Thermodynamics, Second law of, 145

Three periods of progress, 3
Tin plating, 271
Tilden, 146, 298
Titanium, 278, 308
TNT, 19, 21, 84, 299
Trinitrotoluol, 19, 21, 84, 299
Tropics, value of, 96, 156, 165, 190, 206, 213, 216

Schoop process, 272
Serpek process, 31
Silicon, 249, 253
Smell, sense of, 97, 98, 103, 109

Tungsten, 257, 277, 281, 308

Uranium, 28

Vanadium, 277, 280, 308
Vanillin, 103
Violet perfume, 100
Viscose, 116
Vitamines, 211
Vulcanization, 161

Welding, 256
Welsbach burner, 287–289, 308
Wheat problem, 43, 299
Wood, distillation of, 126, 127
Wood pulp, 112, 120, 303

Ypres, Use of gases at, 221

Zinc plating, 271